THE PENGUIN BOOK OF THE PHYSICAL WORLD

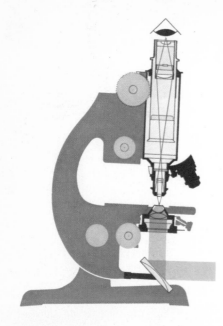

The Penguin Book of

THE PHYSICAL
WORLD

Penguin Books

Penguin Books Ltd., Harmondsworth, Middlesex, England
Penguin Books Inc., 7110 Ambassador Road, Baltimore,
Maryland 21207, U.S.A.
Penguin Books Australia Ltd, Ringwood, Victoria, Australia
Penguin Books Canada Ltd, 41 Steelcase Road West, Markham,
Ontario, Canada
Penguin Books (N.Z.) Ltd., 182–190 Wairau Road, Auckland 10,
New Zealand

First published 1976

Text set in Monophoto Times by Oliver Burridge Filmsetting Ltd,
Crawley, Sussex, England
Printed by Jarrold & Sons Ltd, Norwich, England

Kestrel Books hardcover ISBN 0 7226 5202 X
Penguin Books paperback ISBN 0 14 00 3713 6

THE PENGUIN BOOK OF THE PHYSICAL WORLD

Editorial Consultant: Alan Isaacs, Ph.D., B.Sc., D.I.C., A.C.G.I.

Co-ordinating Editors: Sonya Larkin
Louise Bernbaum

Designer: Bridget Heal

Contents

The world of science is vast and complex. But if we want to understand the rapidly changing world we live in we have to understand what science is and how it works.

Science is also fun. How temperature is calculated, why magnetic filings line up, when our solar system developed – these are pieces of a huge puzzle that scientists have been putting together in theory and experiment for many hundreds of years.

Young people today are exposed to an overwhelming spectrum of knowledge. They tend to acquire a patchwork of facts without an overall framework in which to set the diverse parts. *The Penguin Book of the Physical World* aims to give a bird's-eye view of the story of man and his physical environment. It is arranged so that it can be used as a structured introduction to the world of science, to look up a description of how everyday things work, or just to check a fact.

The book is designed in logical sequence, identified by double-page spread numbers, with each spread forming a separate unit of information, complete with its illustration. Its vocabulary is straightforward and simple but never condescending – if a technical word is the right word, then that is the word used. The diagrams and charts have all been specially made to help the reader to clearly visualize the important ideas behind scientific theories and facts.

The world we live in
1 how science began

Man has always tried to understand the world he lives in. He is distinguished from the rest of the animal kingdom by his understanding of the relationship between causes and their effects – the very basis of science. The evidence provided by the tools and weapons of Stone Age hunters shows that primitive science was already being used by man in his fight for survival, even before the dawn of civilization.

Civilization began with farming. The cultivation and storing of food was one of the greatest revolutions in man's history. From then on man had to learn to live in settled communities. He also had to invent digging sticks, flint sickles, querns for grinding corn and a means of bringing water to the crops. Because farming depends on the changing seasons, a way had to be found to mark the passing of time. The first farmers calculated it by the waxing and waning of the moon, and then they counted the days in a solar year. This marked the beginning of mathematics and astronomy.

The more man came to rely on his tools and weapons, the greater became the need for a material stronger than wood and more easily worked than stone. He probably first discovered iron in the form of meteorites. But before he found out how to make iron from its ore he discovered that copper could be made by burning certain greenish rocks. When he learned, some five thousand years ago, to mix copper with tin, the Bronze Age – and the science of metallurgy – had begun.

Each succeeding generation advanced in scientific knowledge, adding to man's growing store. He used this knowledge to better himself by making his environment work for him.

By the 18th century what we now know as science was called natural philosophy. The terms science and scientist were not used until about one hundred years ago. Today, however, our ever-growing store of knowledge is so vast that it must be broken up into orderly parts. That is why there are separate branches of science such as chemistry, physics, zoology and botany.

Modern science is not simply a boring collection of facts and figures. Scientific research is just as creative as composing a piece of music or making a film. But it is important to have some idea of what science is all about in order to appreciate its beauty and excitement.

The world we live in
2 how science grew

Knowledge and understanding of the world does not grow at a steady pace. It grows quickest when man develops new ways of living; when the nomadic hunter gave way to the settled farmer; when the Bronze Age Egyptian was followed by the Iron Age Hittite, and the Hittite yielded to the Greeks and the Romans.

Science stops growing when a society becomes inflexible. That is what happened in ancient Egypt. The Pharaohs and the priests were content to rely on an endless supply of slaves, so there was no incentive to discover new ways of harvesting crops or to invent labour-saving machines for building palaces.

The Greeks were probably the first to separate science from technology. They were more interested in thinking about a general idea or theory than in putting that idea to practical use. Hero, a Greek mathematician, for example, invented the first steam engine. But because there was plenty of slave labour in ancient Greece the engine remained a toy until it was re-invented two thousand years later by the Englishman, Thomas Newcomen. The Greeks and the Romans were perhaps the first people to try to understand things out of sheer

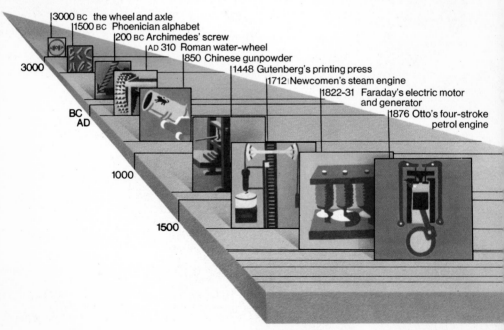

3000 BC the wheel and axle
1500 BC Phoenician alphabet
200 BC Archimedes' screw
AD 310 Roman water-wheel
850 Chinese gunpowder
1448 Gutenberg's printing press
1712 Newcomen's steam engine
1822-31 Faraday's electric motor and generator
1876 Otto's four-stroke petrol engine

3000

BC
AD

1000

1500

curiosity. Many of their theories were somewhat absurd because they saw no need for experiments to try to prove their ideas. But they taught us that to understand something is a valuable exercise in itself, even if it does not have a practical use.

As the Roman Empire crumbled, slavery slowly gave way to the feudal system, and the engineers of the Middle Ages had to learn to manage without slaves. Throughout Europe water-wheels became the great labour-saving machines for milling corn, operating saw-mills and working hammers for crushing metal ores. In southern England alone, by 1086, there were nearly six thousand water-wheels. Horses, too, were more widely used to do the work once done by slaves.

But the greatest increase in scientific knowledge during the Middle Ages came not from Europe but from the Byzantine Empire and the Far East. The Arabs added algebra to the study of mathematics and began the scientific study of chemistry, and the Chinese invented gunpowder. In our own time the pressure of two world wars encouraged the growth of science and technology – nuclear energy, jet propulsion, radar and computers all grew from wartime needs.

landmarks of human inventions

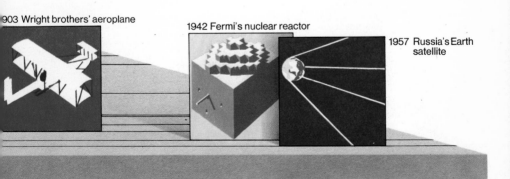

903 Wright brothers' aeroplane

1942 Fermi's nuclear reactor

1957 Russia's Earth satellite

The world we live in
3 · the scientific method

The application of scientific methods to some useful purpose is called technology. Technology depends on a supply of energy. In pre-historic times, and throughout most of history, energy came from man's muscles and those of his slaves or domestic animals.

About seven hundred years ago man first harnessed the power of running water and the power of the wind. Steam power, the first break-through in man-made energy, is scarcely two hundred years old. Only in the last hundred years has electricity been produced on a large scale. The internal combustion engine has been a workable source of energy for about seventy years. Nuclear energy is even younger – the first reactor was developed in 1942. Thus, the real age of science can be said to have begun with the Industrial Revolution between the end of the 18th century and the first quarter of the 19th.

Today most scientific theories begin as they did in ancient Greece, not with technology but as a search for knowledge, which is a way of satisfying natural curiosity. Our method is different from the Greeks', however, in one important respect: they invented theories without testing them. To us a theory has little value unless it can be tested by

The 19th century scientist, Louis Pasteur, making observations in his laboratory.

experiment. This is called the scientific method.

First the scientist decides precisely what aspects of a problem he wants to solve. He then collects all the recorded observations and information about the problem and tries to formulate a hypothesis to explain the information. He may find that some crucial piece of information is missing and this he will attempt to supply by his own experiments. If he is lucky his hypothesis will enable him to make predictions that he can, by further experiments, prove to be true. He will then publish his work in a scientific journal so that scientists all over the world can know about it. At this stage the theory is tested by all the interested members of the scientific community. If anyone can point to a single experiment that does not fit in with the predictions of the hypothesis, the hypothesis must be modified or scrapped.

Once a theory has been accepted it may lie dormant for years before it finds an application in technology. For example, Einstein proposed his special theory of relativity in 1905, but nuclear technology, which depends on this theory, was not developed until almost the end of the Second World War.

Our universe
4   the world of space

It used to be thought that the Earth was at the centre of the universe. Until 1613 it was accepted that the sun revolved around the Earth, but in that year Galileo Galilei published his observations of the heavens which supported the theory of Nicolas Copernicus that the Earth and planets revolve around the sun.

Only in this century have we learned that on the scale of the whole universe, the Earth is a very unimportant object. It is one of nine planets that revolve around the bright star we call the sun. But the sun, on which we depend for nearly all our energy, is itself a very insignificant star in the immensity of space. There are millions of millions of millions of stars like our sun scattered throughout space. We do not know how many of them have planets revolving around them because they are too far away. At these enormous distances we can only identify those objects that emit their own light, like the stars, not those, like the planets and their moons, that reflect it.

The distances between the stars scattered throughout space are enormous. Even in our own solar system, which is the name given to the sun and the planets, distances are very large. If we represented the

the solar system

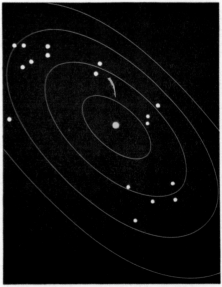

the nearest stars

Earth by a golf ball, the sun would be equivalent to a sphere 4·5 metres across and nearly 0·5 km away. In fact, the distance between the Earth and the sun is about 150 million km. This distance, called an astronomical unit, is sometimes used as a unit of measurement in astronomy. An even larger unit is the light-year, which is the distance travelled by light in one year: it is equal to about 9·5 million million km.

The millions of stars are not scattered evenly throughout the universe; they are collected together in giant clusters (galaxies). The galaxy to which our solar system belongs is called the Milky Way (or sometimes the Galaxy, with a capital G). It is a whirling wheel of some 30 000 million stars. It looks rather like a fried egg in shape, 100 000 light-years across and 10 000 light-years thick at the centre of the yolk. The sun is one small speck in this gigantic whirlpool of stars – situated about where the yolk meets the white – and its nearest neighbour, *alpha-Centauri*, is 40 million million km away. The nearest galaxy to ours is 400 000 times further away even than this and there are thought to be about 1000 million galaxies in the universe.

the solar system in the galaxy

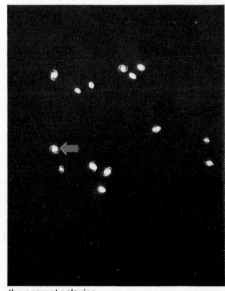

the nearest galaxies

The Earth is a sphere, slightly flattened at the poles – a shape called an oblate spheroid. The diameter at the poles is 12 714 km and it is 43 km greater at the equator. For a sphere of this size it is comparatively smooth and is not much rougher than a billiard ball. The highest mountain, Mt Everest, is less than 10 km above sea level and the deepest ocean is only slightly more than 10 km deep.

The Earth spins on its axis once every 24 hours which makes our day and night. At the same time it orbits the sun once every $365\frac{1}{4}$ days to make our year. Someone living at the equator, even when he is asleep, is travelling through space at 1700 km an hour as the Earth turns on its axis, and at 100 000 km an hour as it travels round the sun.

The Earth's natural satellite is the moon, which orbits the Earth once every $27\frac{1}{4}$ days. The moon is 400 000 km away and has a diameter of 3500 km. Like the planets, it does not produce light of its own but it becomes visible to us on Earth by the light it reflects from the sun. It would take half a million of the brightest full moons to give as much light as we get from the sun. The gravitational pull of the moon on our seas tends to heap up the water which causes the tides. When the gravitational pull of the sun and the moon act in the same direction the high spring tides occur: when the two pulls are at right angles we have the low neap tides.

It is easy to observe shapes on the moon with a telescope, and long ago men gave them names like the Ocean of Storms and the Sea of Tranquillity. But the moon has no atmosphere to make storms and as there is no water there are no seas. When the American astronauts landed on the moon, in July 1969, they confirmed much of what astronomers had expected: an acid, volcanic, rocky surface with no vegetation or life of any kind, and no movement. The rocks which they and later moon explorers brought home are similar to types of rock found on Earth. Because the moon is much lighter than the Earth – the pull of gravity is only one-sixth of what it is on Earth – explorers on the moon feel six times lighter. The moon is a most inhospitable place for man: no air, water or food and the temperature varies from 100°C during the day to −155°C at night.

A view of the Earth from the moon

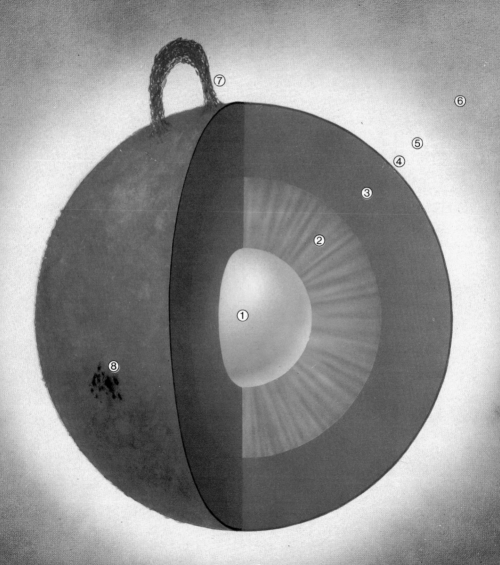

① core 14 000 000°C
② radiation zone
③ convection zone
④ photosphere 6000°C

⑤ chromosphere 35 000°C
⑥ corona 1 000 000°C
⑦ solar flare
⑧ sunspot 4000°C

The sun, our nearest star, is at the centre of our solar system. Compared with other stars it is not very large, but to us it is the biggest and brightest. Films taken when it is eclipsed by the moon show that the sun is surrounded by a fiery atmosphere (the solar corona) which reaches many thousands of miles from the surface. In some places enormous jets (solar flares) gush out even beyond that.

The sun itself is a vast ball of glowing gas. Its outside layer of gas (the photosphere) has a temperature of about 6000°C – about twice as hot as white-hot metal. In places there are blackish spots (sunspots) which sometimes appear in great numbers and then die away. These sunspots are cool areas but they are still hotter (about 4000°C) than the filament of an electric lamp. Sunspots are not yet fully understood but we know that they occur in greater numbers about every eleven years and that they interfere with radio communication.

The sun is over one million km in diameter and 150 million km from the Earth. Light and radiant heat from the sun, travelling at 300 000 km every second, take about eight minutes to reach us. This light and heat are the sources of most of our energy on Earth. Sunlight makes plants grow and without plants no animals could survive. Animals rely on plants for their energy or on other animals, which in turn rely on plants. When plants and animals decay they form fossil fuels, such as coal, oil and natural gas. So all power stations, except those using nuclear energy, rely ultimately on the sun.

The source of the sun's energy used to be a mystery because if it burned like a fire it would have burned itself out long ago. We know now that the sun is like a gigantic hydrogen bomb which continuously changes matter into energy by thermonuclear reactions. In the centre of the sun the temperature is about 14 million degrees C, and at this temperature hydrogen atoms fuse together to make helium atoms, emitting energy as they do so. In this way, when nuclear fusion takes place, some of the sun's substance is changed into energy.

Scientists estimate that the sun will take about 8000 million years to use up its supply of hydrogen atoms. When that happens it will mean the end of life on Earth, but it will still be millions of years more before the sun becomes a completely burned-out star.

Our universe
7 Mercury, Venus and Mars

Unlike the stars, which shine with their own light, the planets of our solar system can only be seen because they reflect the light of the sun. As they are much closer than the stars, we can follow their movements across the sky.

The solar system stretches from the sun, at its centre, to Pluto, the most remote planet, a distance of nearly 6000 million km. Astronomers call Mercury, Venus, the Earth and Mars the inner planets and Jupiter, Saturn, Uranus, Neptune and Pluto the outer planets. Nearest to the sun is Mercury, the smallest of the planets. It revolves very slowly on its axis, taking about 59 Earth days to make one revolution. This creates a very large difference in temperature between the side facing the sun and the dark side. The side facing the sun has a temperature of about 395°C; there is no atmosphere and the surface looks very much like that of the moon.

The brightest object in the sky, apart from the sun and the moon, is Venus. It is about the size of the Earth, with an atmosphere of brilliant white clouds, which prevents its surface from being seen clearly. There is no evidence of oxygen, nitrogen or water vapour in

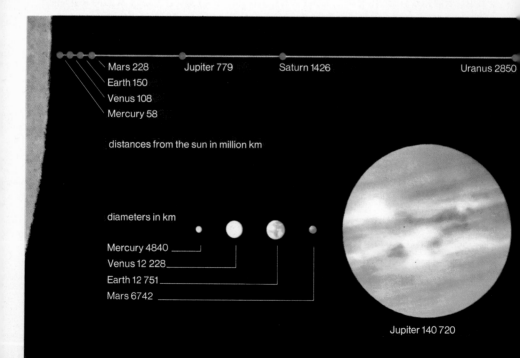

Mars 228 Jupiter 779 Saturn 1426 Uranus 2850
Earth 150
Venus 108
Mercury 58

distances from the sun in million km

diameters in km

Mercury 4840
Venus 12 228
Earth 12 751
Mars 6742

Jupiter 140 720

its atmosphere, although there is plenty of carbon dioxide. Plants and animals such as those which live on Earth could not survive on Venus because the temperature under the clouds is probably about 475°C.

Mars is about half the size of the Earth and, like Venus, has an atmosphere of carbon dioxide. There are dark unaccountable markings on the surface of the planet which seem to change with the period of the Martian year. Early observers thought that they saw long straight lines which they called 'canals'. The photographs sent back in 1969 and 1971 by the Mariner spacecrafts do not reveal any such markings, but show a surface like that of the moon, dotted with huge craters. At the poles there are large white ice-caps: these are probably not frozen water but frozen carbon dioxide. It seems very improbable that life, even of a very elementary kind like lichens, could exist on Mars. Mars has two tiny moons, Phobos (about 19 km in diameter) and Deimos (about 10 km in diameter).

Now that man has reached the moon it will not be long before he takes another giant stride through space to the planets. The first to be visited will be Mars.

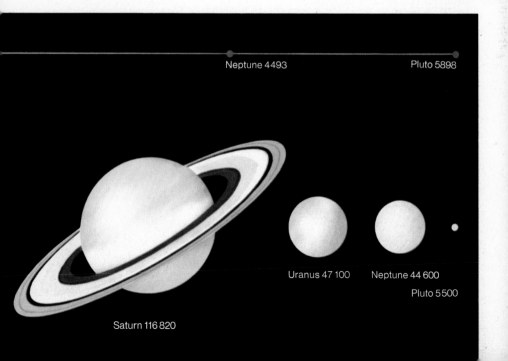

Neptune 4493   Pluto 5898

Uranus 47 100   Neptune 44 600

Pluto 5500

Saturn 116 820

Our universe
8   Jupiter, Saturn, Uranus, Neptune and Pluto

Jupiter is the largest of the planets, nearly 143 000 km in diameter, and weighs more than all the other planets put together. It takes nearly 12 years for Jupiter to circle the sun once. It rotates on its axis faster than any other planet and its day is only 10 hours long. The speed at its equator is some 43 000 km per hour. This high speed of rotation causes a bulging at the equator and a flattening at the poles. There is a deep layer of atmospheric gases, probably made up of ammonia and methane: when seen through a telescope from Earth this deep atmosphere appears to be composed of coloured bands. Jupiter has 12 moons, two of which, Ganymede and Callisto, are larger than our own moon.

Saturn, the second largest planet (about 119 000 km in diameter), is the most extraordinary of the planets because it is surrounded by a system of rings. For a long time the rings puzzled astronomers, but they are now known to consist of millions of separate solid particles revolving independently around Saturn like millions of tiny moons. Evidence for this is that the inner edge of the rings is moving faster than the outer edge and in some places stars can be seen shining

Jupiter is the largest of the planets

through them. In addition to the rings, 10 satellite moons of various sizes revolve around Saturn.

The five planets nearest to the sun have been known from ancient times. They are all bright and can be easily seen with the naked eye. But planets further away were not observed until telescopes were invented. Uranus was discovered in 1781 by the great astronomer Sir William Herschel, Neptune was identified in 1846 and Pluto in 1930. One very odd thing about Uranus is that the axis on which it rotates is in almost the same place as its orbit around the sun which makes it look as if Uranus is spinning around on its side. Not very much is known about Pluto but its orbit has been mapped and it looks as if for part of its path it comes nearer to the sun than Neptune. The paths of all the other planets are quite separate. This suggests that Pluto may at one time have been a satellite which escaped from the gravitational pull of Neptune and began to orbit the sun.

Between the orbits of Mars and Jupiter there is a belt of small bodies, called the asteroids or minor planets. Over 1600 of these bodies have been observed and there are probably thousands more.

Saturn, the second largest planet, showing its system of rings

The Egyptians and Babylonians had an elementary knowledge of the stars and the planets, but they still believed that the Earth was flat and the centre of the universe. The Greek philosopher, Aristotle, explained that the Earth was round, and Ptolemy taught that the sun, the planets and the stars all moved around the Earth. This is what men believed for the next 1500 years.

In 1543 Copernicus, a Polish scholar, suggested that the sun was at the true centre of the universe and that the Earth and the other planets moved around it. But if the Earth was whirling around the sun, how did people stay on it? And what kept the planets in their place? The first man to try to answer these questions was the Dane, Tycho Brahe, who, with no telescope to help him, carefully recorded the movements of the planets. Then, in Italy, Galileo, the first great astronomer to use a telescope, watched the phases of Venus which could only be explained by using Copernicus' hypothesis. In the early 17th century in Germany, Johann Kepler worked out

Top: *Ptolemy*. Centre: *Copernicus*. Below: *Tycho Brahe*

three simple laws to explain why the planets move in elliptical orbits. Then, in England, Isaac Newton introduced the idea of universal gravitation. Newton's brilliant idea was that there is a force of attraction (gravitational force) between all bodies. When he saw an apple falling from a tree he realized that it fell towards and not away from the Earth, because there was a force attracting it. This force of attraction increases with the masses of the bodies but decreases in proportion to the square of the distance between them. We call the force that pulls a body towards the Earth the force of gravity – it is equal to the weight of the body – and it is this force that stops us from falling off the Earth. Newton also saw that the force that pulled the apple to the Earth is the same as the force that holds the moon in its orbit around the Earth, and the planets in their orbits around the sun. For an orbit to be stable this gravitational force has to exactly equal the centrifugal force that is produced when the body rotates about the centre of its orbit.

Top: *Galileo Galilei*. Centre: *Johann Kepler*. Below: *Isaac Newton*

The stars in the northern hemisphere appear to circle around the Pole Star but this is because the Earth is itself rotating. The farmers in Mesopotamia used the position of certain stars to decide when to sow their seeds, and travellers used the stars to guide them. Early observers grouped some stars together into patterns such as the outlines of men and animals. We still call these groups, or constellations, by their old names, such as the Great Bear and the Southern Cross. The stars are so distant that we cannot see if they have planets revolving around them.

Although the stars are like landmarks in the sky, they do not always stay the same; they pass through various stages of development called stellar evolution. A star is born when a mass of hydrogen gas collects in one area of space. The atoms of the gas are attracted to one another by the gravitational forces between them. A great ball of hydrogen forms, and the gravitational forces compress the centre of the ball to such an extent that the temperature rises to millions of degrees. At these tem-

Trifid nebula in Sagittarius

peratures thermonuclear reactions begin, converting hydrogen into helium with the release of enormous quantities of energy, some of which is the starlight that makes the star visible. We call the stars which remain in this state for millions of years main-sequence stars.

When a star has consumed 10% of its hydrogen it expands into an enormous red mass, and is then called a Red Giant. In this state, stars consume hydrogen at a greater rate and some of the helium is converted into the heavier elements. At this stage the star may explode and throw out some of the heavy elements into space. This rare event, a supernova, has been observed only twice, in 1054 and in 1604. A White Dwarf, the star remaining after a supernova explosion, is much smaller than a Red Giant. It is thought that bodies such as the Earth and the planets, which are made of heavy elements, are products of a supernova explosion that have been attracted into orbits around the sun (itself a main-sequence star). After all its hydrogen is gone, a star finally dies and becomes a neutron star.

Pleiades in Taurus

Our universe
11   telescopes and spectroscopes

Less than four hundred years ago Galileo used a telescope, one of the first ever made, to observe the phases of Venus and the four moons of Jupiter. Galileo's telescope was the simplest kind of instrument, a refracting telescope in which the light is collected by a lens. Light from the star or planet passes through the light-collecting objective lens and the image formed is observed through the eyepiece. However, a refracting telescope needs a large lens and it is difficult to make such a lens without flaws – a very large lens is so heavy that it tends to sag in the middle, causing distortion. The biggest successful refracting telescope, with a lens diameter of 40 inches, was built in 1897. Earlier, Newton realized that the difficulties of the refractor lens could be overcome by using a concave mirror to replace the objective lens. Mirrors, too, have their difficulties, but the biggest telescopes today are reflectors. The United States Palomar Observatory has a giant 200-inch reflector instrument.

One disadvantage of optical astronomy is that the light from distant objects has to pass through the Earth's atmosphere; clouds are also a great nuisance to astronomers. These problems have been

Newton's first reflecting telescope was built in 1668

solved by mounting television cameras with telescopic lenses onto artificial satellites, which obtain excellent photographs of some of the planets. Although telescopes can show the planets in considerable detail, a star is too far away to be shown as more than a point of light. Astronomers use a spectroscope which splits the starlight into its components – the band of component colours is called a spectrum. Because each element has its own recognizable spectrum it is possible to tell which elements are present in the stars. Spectroscopes can classify stars according to their colour and temperature.

The newest tool of astronomy is the radio telescope. In 1931, a radio operator took the first steps in radio astronomy when he tried to find out where radio noise – the background hiss you sometimes hear – comes from. Radio telescopes pick up radio waves that come from stars and other bodies in space and enable us to make maps on which the sources of radio waves are plotted. Many of these radio sources can be seen as stars through optical telescopes. Radio telescopes are also used in tracking artificial satellites. The most famous radio telescope is the 250-foot dish at Jodrell Bank in England.

The radio telescope at Jodrell Bank near Manchester, England

Until the middle of the 1950s all our knowledge of the solar system and outer space came to us either by light waves or radio waves. It was like having two windows, one for optical telescopes and spectroscopes and the other for radio telescopes. It seemed to be only a dream of science fiction that man could actually leave the Earth's atmosphere and explore the outermost depths of space.

However, in 1957 Sputnik I (the first artificial Earth satellite) was launched by the Russians and this was followed in less than a year by the American Explorer I. The basic principle of launching a satellite is simply to attach it to the last stage of a multi-stage rocket and accelerate it through the Earth's atmosphere up to 40 000 km per hour. This is the velocity (called the escape velocity) needed to escape from the gravitational pull of the Earth. Once in orbit in space the satellite needs no more power: it will continue to rotate in its orbit like a natural satellite, because in the absence of an atmosphere there are no friction forces to oppose its motion.

Astronomers now have a great number of artificial satellites working for them. O.S.O. (Orbiting Solar Observatory) satellites, for

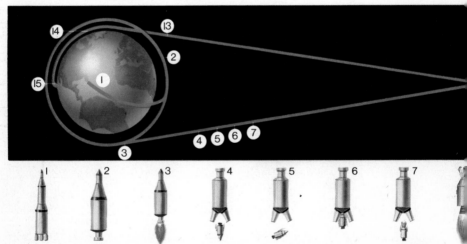

flight of Apollo 11 **1** lift-off from Cape Kennedy **2** second stage falls away and craft goes into Earth orbit **3** rockets fired to put craft out of orbit and into moon trajectory **4** CSM (Command Service Module) detaches from Saturn third stage **5** CSM turns to link with LM (Lunar Module) **6** LM and CSM dock **7** third stage jettisoned **8** craft enters moon orbit and two astronauts enter LM **9** LM separates from CSM

instance, are making continuous observations of the sun. Others are studying the composition, pressure and temperature of the upper atmosphere. Tiros and Nimbus are working for the weather stations collecting information on storms and hurricanes. In 1973 the Americans launched their Skylab satellite, an enormous space laboratory in which two separate crews were able to live and carry out experiments for long periods (59 days in the case of the second crew). The satellites of most practical use to us are the intercontinental communication satellites like Echo, Telstar and Early Bird that make long-distance television transmissions possible. They are also used for intercontinental telephone calls instead of radio.

From the time that Yuri Gagarin (1961) and John Glenn (1962) made the first manned space flights, it took less than 10 years for Apollo 11 to land astronauts on the moon. Yet the moon is only one tiny part of our solar system: vast expanses of space lie beyond, waiting to be explored by man. The moon is undoubtedly our first stepping-stone into deeper space – already a Mars landing is planned for the 1980s and exploratory probes have been sent to other planets.

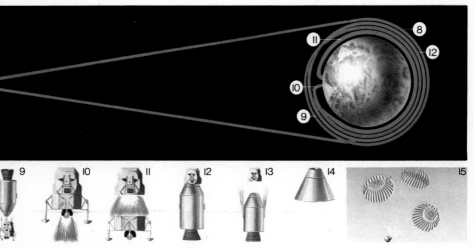

10 LM descends firing retro rockets **11** LM leaves moon and enters orbit leaving descent stage behind
12 LM joins CSM and both go into Earth trajectory **13** CSM leaves LM and enters atmosphere **14** CSM
begins descent **15** parachutes reduce descent speed to 25 mph

Our universe
13   how it began

On a clear night, with the naked eye, you can see about 6000 stars. With modern optical and radio telescopes the horizons of the universe are being swept further back into space and time and it is now thought that the universe contains some thousand, million, million million stars.

Optical telescopes show that these stars are not evenly spread in space, but are collected into giant clusters (galaxies). Our sun is one of the million million stars that make up our local galaxy, the Milky Way. Radio telescopes reveal some very strange objects besides the ordinary stars; the quasars, which emit fantastic quantities of energy as radio waves, and the pulsars which emit their radiation in short regular pulses like the rotating beam of a lighthouse.

We have only one clue as to how it all began. When starlight is passed through a spectroscope the bright spectral lines are displaced towards the red end of the spectrum. This red-shift, due to the Doppler effect, tells us that the galaxies are rushing into space away from our local galaxy, and the further away they are the faster they are receding. One explanation for this is that all matter in the universe started as a single, super-dense mass which exploded, flinging lumps of matter into space like fragments from a bomb. The recession of the galaxies is a continuation of this process. But this 'big bang' theory tells us nothing about the origin of the super-dense agglomeration. Who created it? Why did it explode? Where did the original matter and energy come from?

A rival explanation suggests that the universe did not have a beginning, but has always been there. According to this 'steady state' theory, matter is being made all the time to replace the matter lost by the receding galaxies. The creation of only one hydrogen atom every year in every four cubic km of space would be sufficient to do this. As this rate of creation is far too slow for us to detect in the vastness of the universe, we cannot test this theory directly. In fact, we do not yet know if either of the theories is correct. Perhaps the universe passes through alternating stages of expansion and contraction. Our new radio telescopes are probing deeper into space for the answer.

Great galaxy in Andromeda

The study of the Earth – what it is made of, its tides and weather – has several branches, all known as the Earth sciences. Geology (rocks), meteorology (weather), seismology (earthquakes) and oceanography (oceans) are the principal Earth sciences. Geology, perhaps the fundamental science, began when early man used rocks to make tools, and then later searched for minerals that would yield copper and tin. But the scientific study of geology only began two hundred years ago when William Smith, an engineer, made the first map of England showing the locations of the various kinds of rocks. Geologists use the word 'rock' to include all the materials that make up the Earth's crust, so sand and clay are included. They classify rocks in three groups: igneous, sedimentary and metamorphic.

All rocks were originally formed during the period when the Earth was molten and then cooled down. A great deal of this fire-made (igneous) rock still exists in its original state in many different forms according to the way the molten rock (magma) solidified. A common variety is basalt, a dark-coloured rock formed from the lava of an erupting volcano. In the Giant's Causeway in Northern Ireland, the lava flow of millions of years ago set into great six-sided columns of rock 40 cm wide. In other places lava cooled off with bubbles of gas inside to form pumice stone. Igneous rocks do not contain fossils.

Hard as it is, igneous rock breaks up (weathers) into grains of sand in the course of millions of years by the action of wind and water, sun and ice. Some of this sand settled at the bottom of the sea and over the ages became pressed into hard rock to form layers (strata) of varying thickness and composition of sedimentary rocks, such as sandstone and limestone. Sometimes sedimentary rocks contain fossil remains of plants and animals. In fact limestone largely consists of the skeleton remains of millions of tiny sea creatures.

Metamorphic (Greek for 'changed in form') rocks have been changed by great heat or pressure. For example, clay can be changed to slate by pressure, and limestone into marble by heat. Most of the soil that covers the surface of the Earth is made of all kinds of finely broken rock, mixed with humus, the remains of plants and animals.

The Giant's Causeway in Northern Ireland is a famous example of basalt stone

The Earth sciences
15 our planet's structure

The part of the Earth we know best, the land and sea, is only a thin crust covering the Earth like the skin of a very slightly wrinkled apple. This crust is only about 10 km thick under the oceans, and about 30 km thick under land, a very thin skin compared with the 12 760 km diameter of the Earth. This crust consists of 47% oxygen, 28% silicon (sand consists largely of silicon dioxide, SiO_2), 8% aluminium, 4·5% iron, 3·5% calcium, 2·5% sodium, 2·5% potassium and 2·2% magnesium. All the other elements together only constitute 1·8% of the Earth's crust. Some mines go down about three km and even the deepest borehole, drilled in search of gas or oil, is not much deeper. So scientists have no direct knowledge of the interior of the Earth, and they have to rely on such indirect information as the behaviour of volcanoes, earthquake waves and the vibrations from man-made explosions as they travel through the Earth.

At the centre of the Earth geologists believe there is a solid core some 2800 km across, made of iron and nickel. The fact that the Earth acts like a magnet supports this theory. Surrounding this hard core is probably a region of molten iron and nickel about 2000 km deep and at a temperature of 5000°C. Above this there is a solid region, called the mantle, which is about 3000 km deep, made of heavy rock.

Between the mantle and the crust is a boundary surface resembling a second skin. It is called the Mohorovicic discontinuity (often called the Moho) after a Yugoslavian scientist, Andrija Mohorovicic, who was able to deduce its existence in 1909 by studying earthquakes. The composition of the Moho is not known for certain, but it is thought that within it there is a gradual change in the type of rock – from the basic rock of the crust to the heavy rock of the mantle.

We cannot be certain what the mantle rock is like until we can drill through the crust of the Moho and bring up a sample. During the 1960s a project for boring through the ocean floor was started at a site near Honolulu where the Moho is estimated to be only some 10 km below the crust. It was abandoned because it was costing too much. It is strange to think that we have been able to obtain a sample of rock from the moon's surface more easily than one from a few kilometres below the surface of the Earth.

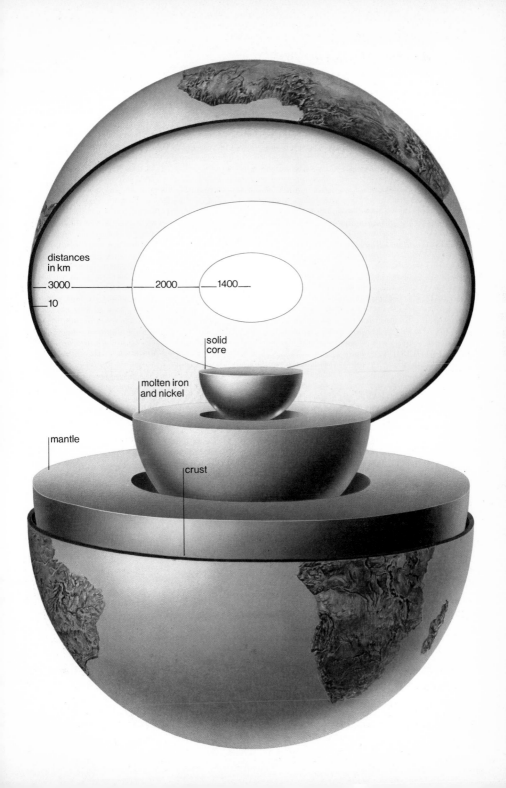

distances
in km
3000
10
2000
1400

solid
core

molten iron
and nickel

mantle

crust

The Earth sciences
16   our planet's history

If you make a stack of newspapers day after day, at the end of the week Monday's paper will be at the bottom, and Sunday's at the top. Every newspaper is dated, but even if it were not, you could guess its date from its position in the pile. It is just like this with layers of sedimentary rock where the deepest layers are the oldest.

As each layer of sedimentary rock was formed, the plants and animals which were alive at that time became buried in it when they died. Usually they simply rotted away without any trace, but occasionally fossil skeletons have survived for millions of years. It is from these fossils that we learn about extinct plants, such as the giant tree ferns, and extinct animals, such as the great reptiles. It is rather like reading in one of the newspapers, picked out of our pile, about what happened on a particular day. We can tell from rocks that dinosaurs lived about two hundred million years ago, and early fishes five hundred million.

Fossils also give us clues about the climate when the fossilized organism was alive. For example, rocks in Greenland contain fossils of plants that can live only in a warm climate; so we can conclude

four thousand million years of the Earth's history

the first green algae
the first soft-bodied animals
the first shellfish
the first vertebrates
the first land plants and animals
the age of reptiles
the age of mammals
the age of man

growth of life

that these northern regions must once have been warmer.

Fossils also give us other clues about changes to the Earth since they were laid down. Those of sea animals found in mountain ranges, for example, show that the areas which are mountains today were once probably under the sea. This evidence suggests that great earth movements must have taken place. Indeed, there is other evidence that on some occasions these movements were so violent that even the order of the rock layers was upset, just as if the order of our pile of papers had been upset. In the Grand Canyon of Arizona, the river has cut a gorge $1\frac{1}{2}$ km deep so that the layers of rock (strata) built up over three hundred million years are clearly visible.

Rocks containing fossils help us to trace the Earth's history back six hundred million years. The older igneous rocks, which contain no fossils, cannot be used in this way, but scientists can calculate their age by testing the radioactive materials they contain: as radio-isotopes decay they form stable products. It is the ratio of active to stable material that provides an age clue from which it is estimated that the Earth started to cool about four thousand million years ago.

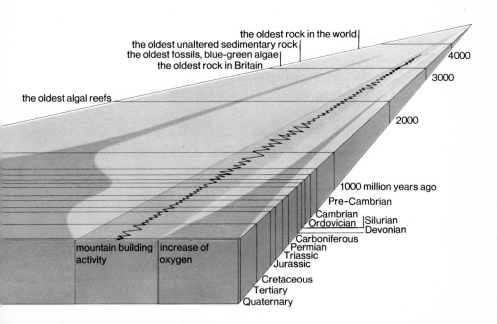

The surface of the Earth is constantly changing. Some changes happen suddenly, such as earthquakes, but most changes take hundreds and thousands of years. As the Earth cooled, its surface shrank and wrinkled, with enormous forces bending and breaking the rock. Geologists call the bends folds, and breaks in the rock are called faults. If you push the two sides of a tablecloth together, its rumpled surface will show how the Earth's crust formed mountains and valleys: these are the folds. You can visualize the formation of faults if you have seen ice breaking up in a stream and watched some of the pieces being forced so that they slide over each other.

Great earth movements have taken place at different periods of geological time, so that mountain ranges in different places are not all the same age. The mountains of Scotland and Wales are about four hundred million years old. When they were young they probably towered twice as high as the Himalayas do today, but ice, water and wind have slowly worn them down. Geologists call this wearing away erosion. The Himalayas are being eroded too, but they are very new mountains, like the Alps and the mountain ranges of Indonesia. These ranges are only forty to fifty million years old.

More gradual changes (erosion) are going on all the time. The sea constantly wears away the land and changes the coastline. It undermines cliffs, makes caves and washes sand from one place to another. Rivers cut gorges, rain swills away soil into streams, and rivers, and glaciers, like great sanding machines, rub away the sides of mountains.

The weather, too, acts on rocks and soil, splitting, breaking and wearing them away. When water finds its way into cracks and then freezes, it expands and can shatter the hardest rock. The hot midday sun and the cold night air create opposing strains in lumps of rock until they crack like a glass which has been heated too rapidly. The wind which drives clouds of sand over the rocks smoothes them in the same way as a house-painter smoothes old surfaces by rubbing them down with sandpaper. Even rainwater, which is specially pure and soft, gradually eats its way into rocks because the carbon dioxide dissolved in it makes it slightly acid.

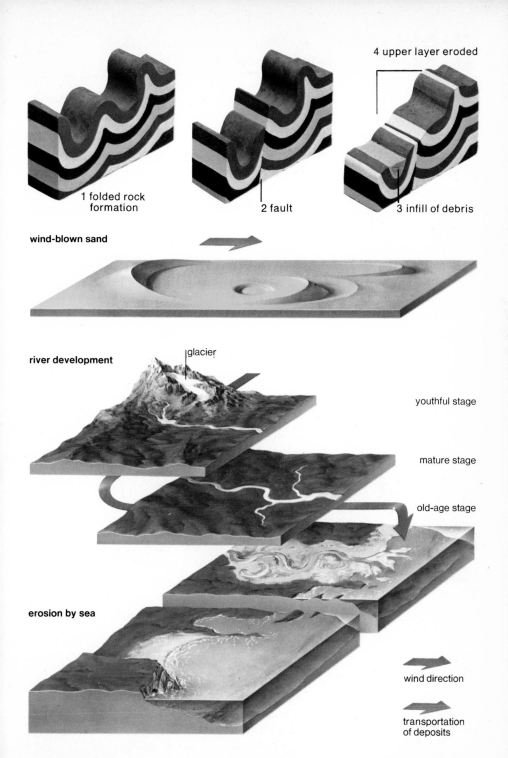

4 upper layer eroded

1 folded rock
formation

2 fault

3 infill of debris

wind-blown sand

river development

glacier

youthful stage

mature stage

old-age stage

erosion by sea

wind direction

transportation
of deposits

The Earth sciences
18 minerals and metals

When we refer to the mineral wealth of a country we usually mean the metals, such as iron and copper, that it produces. But there are other substances – sand and china clay, for example – which are also minerals; they, too, are valuable because they are used to make bricks, concrete and pottery. Coal and oil are also often called minerals, but in fact they are not true minerals because coal consists of the remains of prehistoric forests, and oil of the remains of prehistoric animals. True minerals do not have an animal or a vegetable origin, but are chemical substances that occur in rocks. There are hundreds of different minerals, but only about twenty or so common ones.

Few metals in the Earth's crust are found in a pure state. Often they are combined with oxygen to form oxides, or with sulphur to form sulphides. By far the most widespread chemical elements in the Earth's crust are oxygen and silicon. Although oxygen exists in the air as a gas uncombined with other elements, in minerals it is always combined. Sand, for example, consists of oxygen combined with silicon, and the chemical name of this compound is silica. Metal compounds, like oxides and sulphides, are usually found mixed with

galena   cinnabar crystals   chromite   limonite

bauxite   sulphur crystal   iron pyrite (fool's gold)   magnetite

sphalerite   pitchblend ore   quartz crystal   corundum

sand, clay or limestone: these mixtures are called ores. The important metals such as iron, copper, lead, tin and aluminium are extracted from metal ores. About nine-tenths of all the metal used today is iron or steel. Steel is almost all iron, but it contains about 1% of carbon and often traces of other elements; it is these trace elements that make steel so much tougher than pure iron. There is about ten million times more iron than gold in the Earth's crust. This gives gold a great scarcity value, which accounts for its use in commerce. Gold is unusual because it is sometimes found practically pure in great lumps called nuggets. About a hundred years ago a nugget weighing over 100 kg was found in Australia. But most gold is collected by a costly process of crushing rocks to extract their tiny content of gold.

Diamonds are the hardest of all stones which is why they are used in industry for cutting, drilling and engraving. Many gems owe their beautiful colours to traces of impurities. Ruby, topaz, emerald, sapphire and amethyst are all oxides of aluminium with traces of different metal oxides. A ruby is red because it contains traces of chromium oxide, and traces of cobalt oxide make a sapphire blue.

amazonite

red and white
banded agate

turquoise howlite   lace agate

turritella agate   rhodocrosite,

adventurine   laguana agate   gold tigereye   unakite

The Earth sciences
19   rivers, seas and oceans

Nearly three-quarters of the Earth is covered with water. Water heats up more slowly than land, but once it has become warm it takes longer to cool down. If the Earth's surface were entirely land, the temperature at night would fall quite quickly and night would be much colder than day, as it is on the moon. This does indeed happen in inland deserts, hundreds of miles from any sea. The climate of the continents, especially in the temperate zones, is very much affected by the oceans around them. The areas close to the sea have a 'maritime climate', with rather cool summers and warm winters. The interiors, far from the sea, have a 'continental climate' with extremely hot summers and cold winters.

Rain comes from the evaporation of rivers, seas and lakes. Even after heavy rain, the pavements in a city do not take long to dry because the rainwater evaporates into the air. On a warm dry day it evaporates very rapidly, as warm air can absorb more moisture than cold air. But at any particular temperature, the atmosphere can hold only a certain maximum amount of water vapour. The air is then saturated, like a sponge that cannot hold any more water. The lower

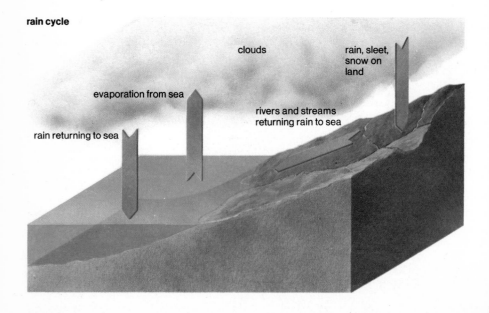

rain cycle

clouds

rain, sleet, snow on land

evaporation from sea

rivers and streams returning rain to sea

rain returning to sea

the temperature, the less water vapour is required to saturate the air.

All over the surface of the Earth, millions of tons of water are evaporating every second, condensing in the air into drops so small that it takes thousands of them to make a single raindrop. It is these tiny droplets that make clouds. When clouds roll in from the sea over the warmer land, they are forced to rise and become cooler in the colder upper atmosphere. As the air cools down it may pass through its saturation point and then some of its water vapour turns to rain. Day in, day out, the same water circulates between the air and the land: rivers evaporate to make clouds, clouds make rain, rain makes rivers which in turn run into the sea. This is called the rain cycle.

The shores around the oceans slope gradually to a depth of about 150 metres. Beyond this continental shelf there is a steep slope down to the deep sea plains some three to 13 km below. In 1960 the Mariana Trench in the western Pacific Ocean which is thought to be the deepest depression on the Earth's surface was explored by oceanographers in a bathyscaphé. The Mariana Trench is deep enough to cover Mt Everest and still leave its tip 1·6 km under water.

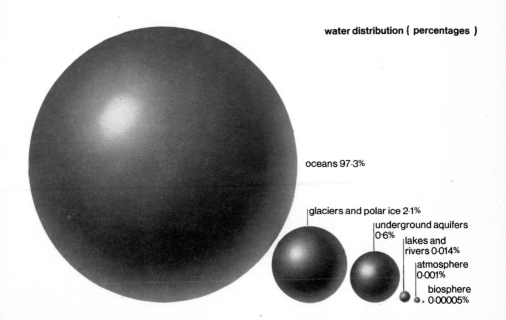

water distribution (percentages)

oceans 97·3%

glaciers and polar ice 2·1%

underground aquifers 0·6%

lakes and rivers 0·014%

atmosphere 0·001%

biosphere 0·00005%

The Earth sciences
20   earthquakes and volcanoes

There is an earthquake somewhere almost every day. It may only be
a slight tremor, it may be a whole series of violent upheavals such as
those in Chile in 1960, or it may be a single shock such as the one that
killed 200 000 people in Japan forty years ago. The Chile earthquakes
started old volcanoes erupting again, and sent giant waves racing
across the Pacific to places as far away as Australia.

Most earthquakes occur in three regions of the Earth: along the
west coasts of North and South America, in a belt across southern
Europe and southern Asia as far as Iran, and in a belt in the Pacific
Ocean including Japan and most of the East Indies. These are the
regions in which mountain-building has been most recent and where
volcanoes as well as earthquakes are likely to occur. The cause of
earthquakes is a movement of rocks some eighty km below the Earth's
surface. This may result from a new crack or fault in the rock layer or
from movement along existing faults. The starting point from which
the fault begins is called the focus and scientists can locate it with an
instrument called the seismograph. By comparing the records of two

*Major earthquake zones appear as thick clusters of dots on this map which covers a 20-year
period of activity*

major earthquake zones

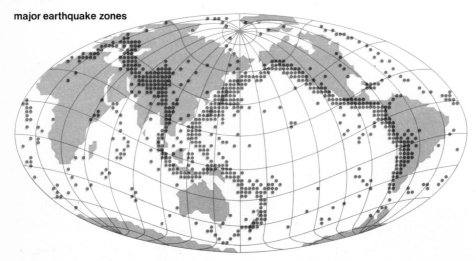

different seismological stations it is possible to fix almost exactly the position of even a small tremor. Seismologists can also tell the distance of the focus below the surface, and as shock waves travel at different speeds through different kinds of rock, they can also locate oil-bearing strata.

A volcano is an opening in the Earth's crust through which molten rock (magma), steam and other gases are forced up. Volcanoes are classified as active, dormant or extinct. An example of an extinct volcano is Crater Lake in Oregon, in the United States, where a great lake fills the crater. Vesuvius is at present dormant but in A.D. 79 it destroyed the Roman cities of Pompeii and Herculaneum. Etna in Sicily is still active. Sometimes a volcano overflows quietly as lava pours out: others explode with terrific force, as Mt Pelee in the West Indies did in 1902 when it destroyed an entire city of 28 000 people.

Geysers, located in volcanic regions, are shooting springs of hot water. The volcanic magma heats the water. Old Faithful, in Yellowstone National Park in the United States, erupts every few days and throws up half a million gallons of water every hour.

Diagrammatic section through an active volcano

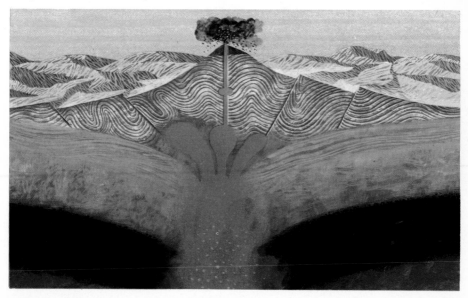

The Earth sciences
21   waves, currents and tides

The ocean basins average between three and five km deep, and their floors are crossed with mountain ranges, valleys, volcanoes and great underwater plains. The deep water near the sea-bed is only about one or two degrees above freezing as, even in the tropics, the sun's rays are unable to penetrate this depth of water.

The waters of the seas and oceans are in constant motion. As well as the waves on the surface, there are currents flowing below like sea-rivers and there are the vast swinging movements of the tides. Waves are mainly caused by wind. If you watch a piece of wood floating on a lake, you can see that although the ripples move it up and down, it scarcely changes its position in relation to the shore. In mid-ocean the difference between the bottom and top of the swell may be more than thirty metres although it is usually no more than half of that.

There are two kinds of ocean currents, one near the surface and the other in the depths. The direction of surface currents differs north and south of the equator. The surface currents flowing towards the poles are narrow and fast but those flowing towards the equator are wide and slow. One of the best-known ocean currents is the Gulf Stream

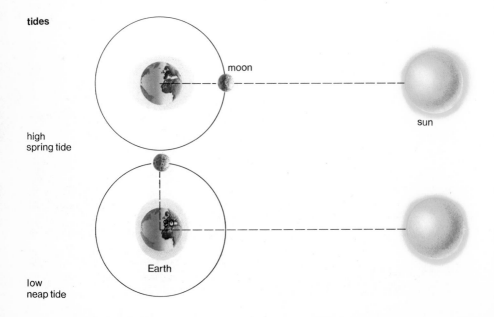

tides

moon

sun

high
spring tide

Earth

low
neap tide

that carries relatively warm water up the east coast of North America and across the Atlantic to the coasts of Europe. Far below the Gulf Stream, moving in the opposite direction, is one of the great deep-water ocean currents. Ocean currents are caused by prevailing winds, by differences in temperature and also by the spin of the Earth.

Tides are caused by the pull of gravity, mainly between the Earth and moon but also, to a lesser extent, between the Earth and sun. The gravitational attraction between the Earth and the moon pulls the waters of the seas towards the part of the Earth nearest to the moon and thus produces a high tide. The moon also pulls the solid earth so that on the side farthest away the effect is to pull the land slightly away from the water, causing a second high tide. The movement of the Earth and moon causes high tides at regular intervals of 12 hours 25 minutes. When the sun and moon are in line on the same side of the Earth, the sun's gravitational attraction is added to that of the moon and creates the exceptionally high tides of spring. When the gravitational forces of the sun and moon are in opposition, with the Earth between the sun and moon, the low neap tides occur.

surface currents

180°
120° 120°
60° 60°
0°
Arctic Circle

Tropic of Cancer

equator

latitude
longitude

the longitude of a point is the angular distance
between the meridian passing through that
point and the zero meridian

Some cities are built with straight streets and avenues crossing one another at right angles, which makes street directing very easy. In New York, for example, it is simple to find the corner of 44th Street and 3rd Avenue. The criss-cross lines on a map, the Earth grid, help in the same way. Chicago is 42°N and 88°W, so it is at the junction where the 42 line running horizontally (latitude) crosses the 88 line running vertically (longitude). 'N' means north of the equator, and 'W' means west of the zero meridian, which is the line of longitude passing through Greenwich Observatory in London.

The idea of an Earth grid started with the Greeks more than 2000 years ago. Lines of latitude are imaginary circles parallel with the equator, running around the Earth. We call the equator 0 and measure the other parallels of latitude from it: the North Pole is 90°N, and the South Pole is 90°S. If we draw parallels of latitude at each degree from 0 to 90 the distance from any one to the next one is about 100 km. But the lines of latitude themselves grow shorter the further they are from the equator.

Lines of longitude are each half of imaginary great circles drawn around the Earth from north to south. They are all the same size, cross each other at the poles and cross the parallels of latitude at right angles. As there is no obvious choice of meridian from which to start counting, in 1884 geographers selected the meridian that runs through Greenwich. Longitude fixes the position of a place east or west of this prime (or zero) meridian. At the equator, the distance of one degree of longitude from its neighbour is about 112 km; moving north or south of the equator, lines of longitude gradually draw together until at the poles there is no distance between them.

As the Earth revolves on its axis once each day, in 24 hours it turns through 360°. In one single hour, therefore, it turns 15°. Since the Earth turns from west to east, when it is noon in Greenwich it is one hour before noon for each 15° to the west, and one hour after noon for each 15° to the east. When it is noon in London, it is only 6 a.m. in New York, while in Calcutta it is already 5.30 p.m. If the country is very large, the time can vary within it: for instance, when it is 7 a.m. in New York, it is only 4 a.m. in San Francisco.

The Earth sciences
23 making maps

Because the Earth is a sphere, a true map of it can only be drawn on a globe. Four hundred years ago, famous voyagers always used a globe. However, it is more convenient to have maps on flat sheets in order to make enlargements of a particular area to any scale. The scale of the map on a globe is determined by the diameter of the globe.

When surveyors are making a map they draw a base line between two chosen points several miles apart and measure angles from either end of the base line to a third point. They can draw this triangle accurately to scale. The two new sides are used as base lines for more triangles, and this can be continued until the whole area has been mapped. In the 1824 survey of Britain, 250 triangles were used with sides varying from 10 to 110 miles. This is called triangulation. It is very useful for making maps that accurately show the distances between towns, the positions of rivers or the shape of a coastline. But, like all maps on flat paper, it would not be quite true as it does not represent the curvature of the Earth. The bigger the area covered, the greater a flat map distorts, so mapmakers use different projections to show curved areas on flat paper. Some projections are good for

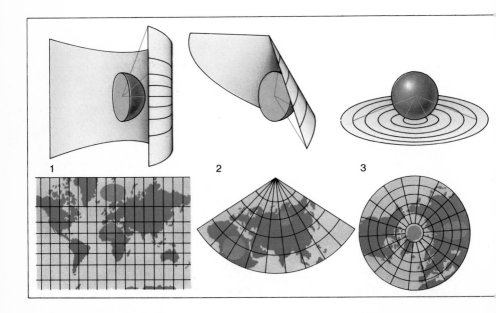

showing the shape of countries, others can show lands near the equator accurately but distort regions near the poles.

In 1569 the Flemish geographer and mathematician, Gerhardus Mercator, projected the globe onto a cylinder. This method makes parallels of latitude all the same length, instead of becoming shorter as they move towards the poles. Imagine a map of the world printed on an orange. If you peel the orange by cutting the skin into small segments and then press them down flat this will show how Mercator made his map. The areas and shapes near the equator are little changed, but further north or south the distortion is exaggerated. For instance, Greenland looks about as big as South America, but actually it is only one-eighth as large. Mercator's projection is useful because a line joining any two places gives a true compass direction. But although the direction is correct, the line on the Mercator map does not give a true distance. The shortest distance between two places on the Earth is always along a great circle. For finding this true short distance navigators such as air navigators have to use both Mercator charts and great-circle maps for plotting their courses.

map projections

1 Mercator
2 conical
3 zenithal or azimuthal
4 interrupted Sanson-Flamsteed

4

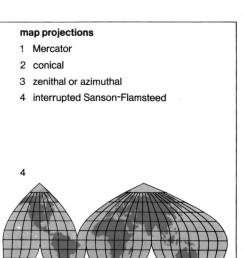

how a surveyor makes a projection

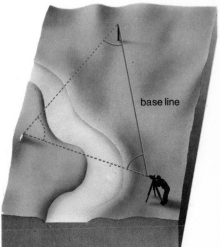

base line

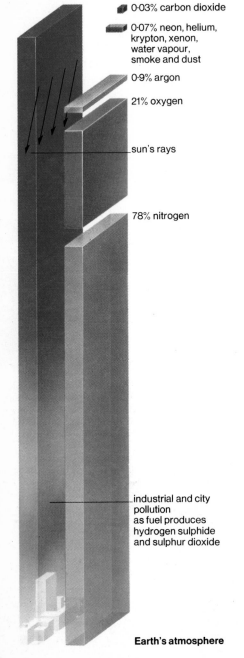

0.03% carbon dioxide

0.07% neon, helium, krypton, xenon, water vapour, smoke and dust

0.9% argon

21% oxygen

sun's rays

78% nitrogen

industrial and city pollution as fuel produces hydrogen sulphide and sulphur dioxide

Earth's atmosphere

The atmosphere
24   the air around us

The Earth is surrounded by the atmosphere which is a mixture of gases. The atmosphere is most dense at ground level and thins out as it stretches upwards until, at about 500 km it is almost a vacuum – about one million times less dense than it is at sea level. Even at half this height above the Earth, the atmosphere is so thin, because there are so few gas molecules, that the air resistance (drag) on an artificial satellite is negligible. The atmosphere provides insulation for the Earth, acting as a sunshade against excessive heat during the day and a blanket to prevent too much cooling off at night. It also filters out harmful radiation from the sun and causes meteorites to burn up by friction before they hit the Earth. The moon, with no atmosphere and thus no friction to protect it, has enormous variations in temperature and is pitted with meteorite craters.

Nine-tenths of all the substance, or mass, of the atmosphere is in the layer from ground up to about 15 km. This lower layer consists mainly of nitrogen (78%) and oxygen (21%). The rest is made up of the rare gas argon (0.9%) and traces of the other rare gases neon, helium, krypton and xenon, which together with varying amounts of water vapour, smoke and dust make up about 0·07%. The atmosphere also contains some 0·03% carbon dioxide. The gases are mixed together, not chemically combined, so that the composition of the air varies slightly in different parts of the world. The main gases, oxygen and nitrogen, are useful chemicals and by compressing and cooling air to nearly $-200°C$ these two constituents can be liquified and stored separately. Oxygen is used in many industrial processes, such as in welding torches for cutting steel, and in hospitals to help patients who are too weak to breathe ordinary air. Nitrogen, combined with other elements, is a valuable fertilizer. Burning and breathing both depend on oxygen. When substances containing carbon, such as coal or oil, burn in air solid soot particles often form with the waste gases. Even a small town with 10 000 people can give off more than five tons of polluting soot and toxic waste gases every day from burning these fuels.

Industrial area in the Midlands of England

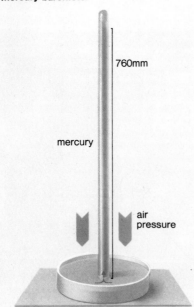

Torricelli's experiment

mercury barometer

760mm

mercury

air
pressure

The atmosphere
25   air pressure

Air is all around us. We cannot see it but we are conscious of it every time we take a deep breath, and we feel it when the wind blows. Although invisible, it has substance and consists of atoms and molecules in the same way as liquids and solids. However, air is about one thousand times less dense than water and one cubic metre of it weighs about $1\frac{1}{4}$ kg. This means that the air in an average-sized single bedroom weighs about as much as a six-year-old child sleeping in it. As air has weight it exerts a pressure on all the surfaces with which it is in contact. At sea level this pressure is equivalent to about one kg on every square centimetre.

It was in 1644 that an Italian, Evangelista Torricelli, devised an experiment to show that air exerts pressure. He sealed a glass tube at one end and filled it with mercury, then he turned it upside down and placed the open end in a dish of mercury. The mercury in the tube fell slightly, but did not all run out. The pressure of the air upon the mercury in the dish supplied the force required to support the 76-cm column of mercury in the tube. The height of the column does not depend on the diameter of the tube but only on the pressure of the air on the mercury in the dish. This became the basis for the mercury barometer. As the pressure of the air varies slightly from time to time, it causes a rise or fall in the level of the mercury. Such changes are useful in forecasting weather: usually, high atmospheric pressure means fair weather and when the barometer falls rain may be expected. Other scientists discovered that the barometer can also estimate height. Since the higher you go the less air is above you, the pressure it exerts is weaker and the mercury falls. For every 100-metre increase in height above sea level the mercury falls about one cm. In 1648 the French scientist, Blaise Pascal, took a barometer up a mountain to demonstrate that at heights above sea level atmospheric pressure falls. Over a hundred years later the first balloonists carried barometers to estimate their altitude. Aircraft altimeters are a special form of barometer (aneroid) to measure height above sea level. An aneroid barometer consists of a metal capsule from which the air has been removed. As the air pressure outside the capsule increases the surface is forced inwards causing the levers to move the pointer.

aneroid barometer

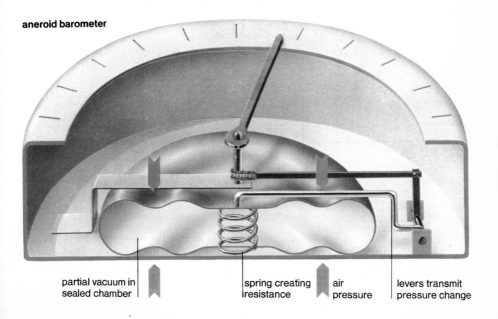

| partial vacuum in sealed chamber | spring creating resistance | air pressure | levers transmit pressure change |

The atmosphere
26 *its three main layers*

The exploration of the upper atmosphere has become possible only in recent times. Earlier climbers discovered that the thin mountain air made them dizzy and sluggish, and the first balloonists in the 19th century to rise 10 km narrowly escaped death from lack of oxygen. In the last quarter of this century pressurized aircraft, unmanned balloons with recording instruments, and satellites launched by rockets have enabled us to find out about the higher regions of the atmosphere.

The air surrounding the Earth consists of three main layers. The lowest, the troposphere, stretches up to about 15 km and contains 90% of the mass of the air. The temperature of the troposphere falls rapidly with increasing height and at its upper limit is about $-60°C$. This upper limit varies slightly from place to place – it is lowest over the poles and highest over the equator. It is the great moving masses of air in the troposphere that mainly determine our weather.

Next comes the tropopause, a dividing layer between the troposphere and the stratosphere, and a pilot can usually tell when he enters the tropopause because the temperature rises a little. The stratosphere itself, the second main layer, consists throughout of very thinned-out gases and it reaches a height of some 50 km. The temperature remains more or less constant in the stratosphere and there are no storms, clouds or haze in this region. Its most striking feature is a belt of ozone, a form of oxygen, that absorbs short wave ultraviolet radiation from the sun. Without the ozone belt life on Earth would be impossible, as this radiation kills living organisms.

Beyond the stratosphere is a belt in which the gases are even thinner and their atoms are electrically charged (ionized) by the sun's radiation. This is the ionosphere which makes long-range radio communications possible because it reflects radio waves back to Earth, which otherwise would travel out into space. But, for the same reason, the ionosphere is an obstacle to radio astronomy because it absorbs much of the radiation emitted by stars. This means that at certain wavelengths the stars cannot be observed by radio telescopes. It is also in the ionosphere, over the North and South Poles, that the beautiful colour effects of aurora borealis and aurora australis occur.

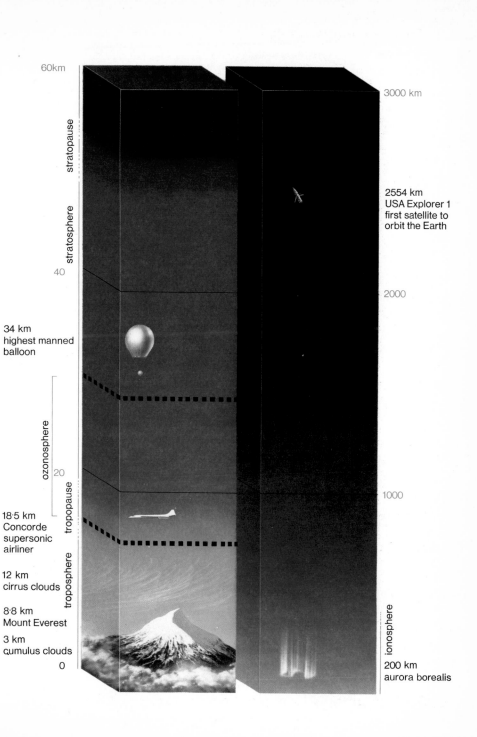

60km

stratopause

3000 km

stratosphere

2554 km
USA Explorer 1
first satellite to
orbit the Earth

40

2000

34 km
highest manned
balloon

ozonosphere

20

1000

18·5 km
Concorde
supersonic
airliner

tropopause

12 km
cirrus clouds

troposphere

8·8 km
Mount Everest

ionosphere

3 km
cumulus clouds

0

200 km
aurora borealis

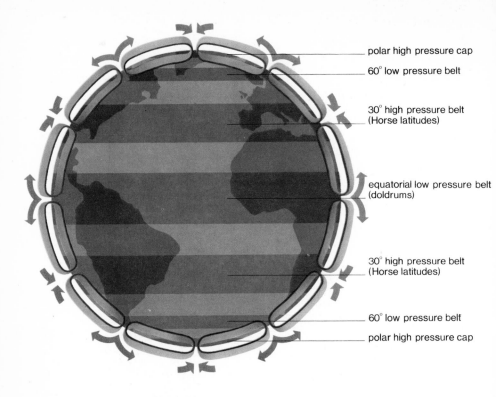

polar high pressure cap

60° low pressure belt

30° high pressure belt
(Horse latitudes)

equatorial low pressure belt
(doldrums)

30° high pressure belt
(Horse latitudes)

60° low pressure belt

polar high pressure cap

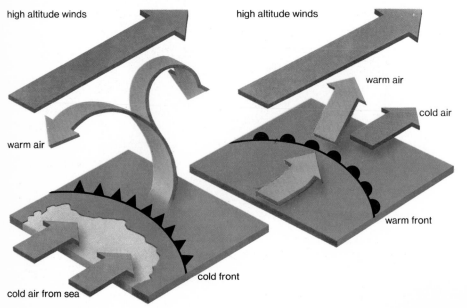

high altitude winds

high altitude winds

warm air

cold air

warm air

cold front

warm front

cold air from sea

Winds are movements of large quantities of air set in motion by differences of temperature between two nearby areas. Winds operate rather in the same way as the air when a radiator heats a room: the air around the radiator expands as it is heated and rises while cooler air, which is heavier and denser, is drawn in to take its place. These movements of air are called convection currents. On the Earth, the heat of the tropics acts like a radiator to heat the air, making it less dense and causing it to rise so that cooler air is drawn in from elsewhere to take its place. As a result, most winds that blow over the Earth move from the cold poles towards the equator. But it is not quite as simple as this. As the warm equatorial air moves towards the poles it gradually cools, so that by about latitude 30° north and south, it sinks and forms new high pressure regions. Also, the rotation of the Earth swings the masses of air around faster at the equator than at the poles. As a result, the winds in the northern hemisphere are deflected to the east and those in the southern hemisphere to the west.

This global pattern of winds is also somewhat modified by the oceans. As the winds blow over oceans and continents local forces come into play, sometimes overriding the global forces. Land masses warm up and cool down more quickly than the sea, so new high and low pressure areas are formed at different times of the day and night in different seasons. These local differences affect not only the direction of the winds, but also their moisture content.

When a mass of warm air overtakes a cold mass, the line at which they meet is called a warm front. The warm air cools as it rises and its water vapour condenses to form rain clouds. A cold front occurs when a mass of cold air overtakes a warm mass, pushing underneath it. Again, the warm air is pushed up and there is often rain, followed by more settled weather as the cold front moves forwards.

The pattern of winds was especially important in the days of sailing ships. Sailors could rely on steady easterly winds on both sides of the equator, which they called trade winds. But along the equator itself there are only light, fitful breezes. To be becalmed there was a sailors' nightmare: they were in the doldrums and the name for this belt of calm weather is still used.

The atmosphere
28 typhoons and hurricanes

You can guess at the speed of a wind by noticing how smoke rises or watching how it sways the branches of trees. The scientific instrument for measuring wind speed is the anemometer which is made of four little metal cups attached, one at each end, to two crossed arms that spin on a shaft. It is important for the safety, especially of ships and aircraft, to describe winds accurately: to do this weathermen and sailors use a scale invented by Admiral Francis Beaufort at the beginning of the 19th century. The Beaufort scale goes from zero, which is dead calm, to 12 and over, which is hurricane. Force 6, in between, is described as strong and it is enough to sway branches and make telegraph wires whistle.

A hurricane is a wind that blows at a speed of more than 120 km per hour. It usually starts over an ocean and may last for several days. These particularly violent storms, common between August and October, develop in the western Atlantic and sweep over the West Indies, the Gulf of Mexico and the east coast of North America. The United States Weather Bureau designates each year's hurricanes, in

This cyclone, swirling over the Pacific Ocean 1200 miles north of Hawaii, was observed by Apollo 9 in March 1969

alphabetical order, with girls' names.

Reports from the weather bureau hurricane hunters who follow the course of every dangerous storm, make it possible to alert areas in its path so that people can take refuge. In 1900, before this warning system was developed, one tremendous hurricane killed over 6000 people. Tropical storms in the China Seas are called typhoons (from the Chinese *tai fung*, big wind). In other parts of the world they are called cyclones. These winds whirl around the centre (eye) of the storm at speeds up to 250 km per hour and the whole wheeling system, which may measure as much as 450 km across, moves forward at about 30 km per hour. Fortunately, if they reach land, storms like this generally die away fairly quickly.

Tornadoes, or whirlwinds, are another dangerous kind of storm, which usually start over land and are accompanied by heavy rain and thunder. They are rarely more than a few hundred metres across, but they whirl so fast that they can devastate areas by uprooting trees and lifting buildings off their foundations.

Tornadoes sometimes move as fast as 800 km per hour, the speed of a revolver bullet. This picture of a tornado in Oklahoma shows the funnel darkened by debris scooped up by the winds

The sun evaporates tremendous quantities of water from the sea and land. Every day there are millions of tons of water carried in the atmosphere either as invisible vapour or as clouds. Warm air can carry far more moisture than cold, so when it cools it can no longer retain the water vapour. It then condenses around the tiny particles always present in the air to form microscopic droplets, as tiny as a thousandth of a millimetre in diameter, that are light enough to float in the sky. These are clouds, of which there are three main types: cirrus (Latin for curl), stratus (layer) and cumulus (heap). The Latin word nimbus is often added to indicate rain clouds. Not all clouds fall into one of these three types; some are mixtures of two types.

Why clouds turn into rain is still a puzzle. Most meteorologists think that in temperate zones rain starts high up at the top of clouds as ice crystals. Some melt and fall as rain and some freeze again after melting and become sleet or snow. When rising warm moisture builds up into the towering cumulus called a thundercloud, the clouds become electrically charged with up to millions of volts. As the electricity discharges from one cloud to another, or from cloud to Earth, the electric current heats the surrounding air to white heat – the lightning flash. Instantly the air expands and it is this expansion that causes the sound waves of a thunderclap. You can calculate your distance from a storm by measuring the time between the flash of the lightning and the thunderclap. As sound travels about three km in one second, for every second between the flash and the clap the storm is three km away. Thunderstorms cause comparatively little damage, but one may release as much energy as several atom bombs, and there are hundreds of such storms around the world every day.

As the air in polar regions is cold, it carries little moisture, so clouds are rare and the fall of rain or snow is slight. In the tropics, however, heavy rain occurs almost every day with great regularity. In the morning the hot sun soon evaporates the moisture on the ground to build up cumulus clouds with a high water vapour content. More moisture is evaporated, and by late afternoon the air is saturated. Suddenly a tremendous downpour, often accompanied by thunder, clears the sky. The stage is set for a repeat performance next day.

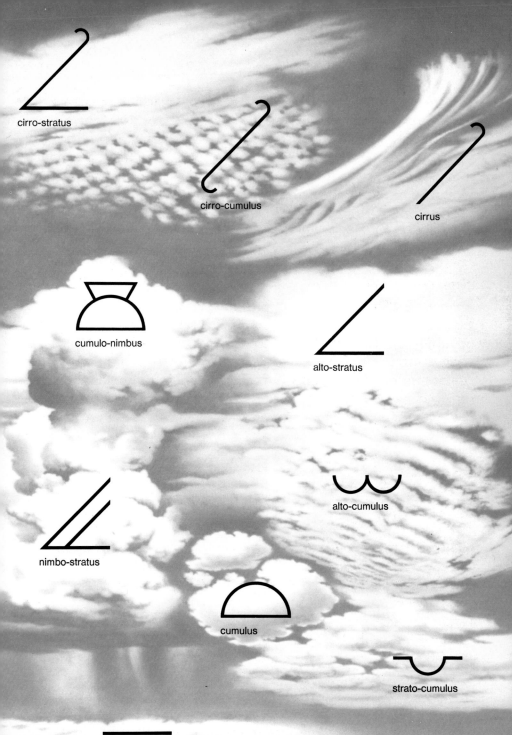

cirro-stratus

cirro-cumulus

cirrus

cumulo-nimbus

alto-stratus

nimbo-stratus

alto-cumulus

cumulus

strato-cumulus

stratus

Weathermen make forecasts by piecing together information from thousands of weather stations throughout the world. The United States Weather Bureau gets detailed reports at least four times every day from over 2000 stations in North America and the West Indies; in Britain reports from amateur weather stations, including many schools, are added to the official reports from the Meteorological Office. Details come in from weather ships and from aircraft on meteorological flights; radiosonde balloons flash data signals from 30 km above ground, and weather satellites with TV cameras transmit weather pictures of great areas of the Earth's surface.

The basic information used in weather forecasting consists of the temperature, pressure and rainfall figures. A continuous temperature record is kept with a thermograph – a special thermometer that operates a moving pen which traces a path on paper wrapped around a revolving drum. A barograph is a similar instrument for making a continuous record of air pressure. The wet and dry bulb thermometer (hygrometer) measures the dryness or humidity in the air. It consists of two thermometers, of which one bulb is surrounded by moistened

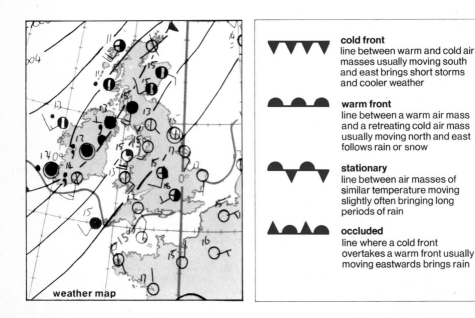

weather map

cold front
line between warm and cold air masses usually moving south and east brings short storms and cooler weather

warm front
line between a warm air mass and a retreating cold air mass usually moving north and east follows rain or snow

stationary
line between air masses of similar temperature moving slightly often bringing long periods of rain

occluded
line where a cold front overtakes a warm front usually moving eastwards brings rain

muslin. This one will give a lower reading than the dry bulb owing to the loss of heat by evaporation. The difference in the two readings depends on the relative humidity of the atmosphere. The simplest instrument is the rain gauge, a funnel standing in a graduated bottle. A sunshine recorder is a glass ball that focuses the sun's rays onto a curved piece of paper and leaves a burn-track as a record.

To make their forecasts meteorologists gather all their information, put it in a computer and then make a weather map. The pressure areas are marked 'low' and 'high', and isobars are lines joining places with the same atmospheric pressure. Barometer readings are marked in millibars and inches (1000 mb = 29·54 ins.). Code marks show the wind direction and its force. Bold lines mark out the warm and cold fronts and the half-circles and spikes on them show the direction in which they are moving. There are different symbols for temperature, rain, snow and sunshine. From all this information meteorologists can give warning of danger zones of gales, storms and general weather conditions. This is extremely important and no aircraft takes off and no ship leaves port without first getting the latest 'met. report'.

weather		cloud		wind	
	mist	◯	clear	◎	calm
	fog	◑	1 okta	◯—	1-2 knots
	drizzle	◔	2	◯—⟍	3-7
	rain	◑	3	◯—╱	8-12
	snow	◑	4	◯—⫽	13-17
	hail	◕	5	◯—⫻	18-22
	shower	◕	6	◯—⫻	23-27
	thunderstorm	◕	7	◯—⫻	28-32
	dust or sandstorm	●	8	◯—⫻	33-37
	glazed frost	⊗	sky obscured	◯—⫻	38-42
			1 okta = ⅛ of the sky	◯—⫻	43-47
				◯—▲	48-52

Egyptian

1984

Babylonian

1984

Roman

MCMLXXXIV

1984

Arabic

1984

Chinese

1984

No one knows when man first began to use language or when he first began to count. But primitive men probably matched the things they wanted to count against something else, such as notches in a stick, or pebbles or knots tied on a cord. As each day passed, for instance, a pebble could be added to a pile. We say now that there is a one-to-one correspondence between the things we count and the things we count them with.

The next step in the development of counting was to use names for numbers. A group of objects could be counted by saying or thinking the names: for instance, we say one, two, three as we count. In the same way it is convenient to use symbols for writing numbers. The ancient Egyptians simply made upright marks from right to left on their papyrus. One was I, two was II and so on.

This works well enough for counting small numbers of things but it is unwieldy for large numbers. One would have to remember too many number names and write down too many marks. To avoid this, a method of grouping was used. The Egyptians grouped numbers in tens. When they reached nine upright marks they used a different symbol, Λ, for 10. Ten 10s were written ℓ.

Many peoples have based their number system on 10, because we have 10 fingers and thumbs. However this is not the only way. The Romans grouped things into fives as well as 10s and their symbol V for five is thought to represent a hand with the fingers together and the thumb outstretched. The symbol X for 10 represents two Vs, one on top of another, with the lower V upside down ($_\Lambda^V$). The Babylonian system involved a base of 60 and although our present system is based on 10, the 60 is preserved in the way we subdivide angles and units of time (60 seconds = one minute, 60 minutes = one hour, and 360° is one complete turn).

The present Hindu-Arabic system originated in India. Like the Egyptians we use a base of 10. But we have different symbols for the numbers one to nine. Tens are indicated by moving the numbers one space to the left. The symbol 0 for zero is used to show an empty space. This is known as a positional notation.

Top: Egyptian tablet lists captives 1400 B.C.
Below: astronomical data recorded in 103 B.C.

The idea of writing numbers using a positional notation and a zero is so familiar that we tend to take it for granted. But, in fact, the invention of this system was almost as important as the invention of the alphabet. Its advantage is that it can be used for calculation as well as for recording numbers. To appreciate the efficiency of our number system, you have only to compare a simple sum as we do it with the same sum written in Roman numerals.

Our Hindu-Arabic system is decimal; that is, the numbers are based on ten. For instance, 13 579 is a short way of writing 1 ten thousand + 3 thousands + 5 hundreds + 7 tens + 9 units. A ten is a group of ten units; a hundred is a group of ten tens; a thousand is ten hundreds. The number is equivalent to $(1 \times 10^4) + (3 \times 10^3) + (5 \times 10^2) + (7 \times 10) + 9$. Here 10^4 stands for $10 \times 10 \times 10 \times 10$ and 10^3 for $10 \times 10 \times 10$. Ten is called the base of the number system.

In the same way that we take the positional notation for granted we also take the base of ten for granted. However, numbers can have any base – if we had six fingers on each hand we might well use a base of 12 – we sometimes do when we talk in dozens and gross. Another

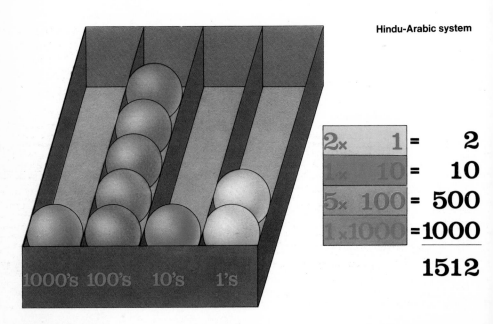

Hindu-Arabic system

$2 \times 1 = 2$
$1 \times 10 = 10$
$5 \times 100 = 500$
$1 \times 1000 = 1000$

1512

1000's 100's 10's 1's

system now in use is the binary notation, which has a base of 2.

Binary numbers are written using only two symbols, 0 and 1. To count in binary, start with 1 for one. Two is written as 10; it is read 'one-oh', not ten. Three is 11, four is 100, five is 101, six is 110, seven is 111, eight is 1000, sixteen is 10 000, and so on. In the binary system a new digit has to be added at two, four, eight, sixteen and so on – not at ten, hundred, thousand, and so on, as in the decimal system. For instance, the binary number 11111 stands for 1 sixteen + 1 eight + 1 four + 1 two + 1 = 31.

Note that four is two twos, eight is two fours, sixteen is two eights, etc. The number can be written $(1 \times 2^4) + (1 \times 2^3) + (1 \times 2^2) + (1 \times 2) + 1$; in this case 2^4 stands for $2 \times 2 \times 2 \times 2$, 2 being the base of the system.

The binary system can also be used for calculations – you have only to remember that $1 + 0 = 1$, $1 + 1 = 10$, $1 \times 0 = 0$ and $1 \times 1 = 1$. However, the numbers used are usually larger than their decimal equivalents (compare 11111 with 31). The importance of binary numbers is that they are used in computers: a flow of current signifies 1, no flow signifies 0. This makes calculations very simple.

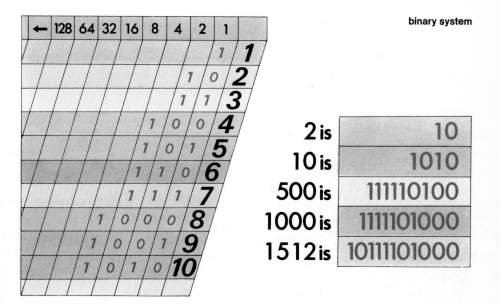

binary system

2 is	10
10 is	1010
500 is	111110100
1000 is	1111101000
1512 is	10111101000

Mathematics
33 calculating machines: abacus to computer

One of the earliest devices for working out calculations was the abacus. It is still used today in some far eastern countries and experts can operate it at a great speed. One type of abacus has two frames. Each bead moved up on the bottom frame represents 'one' and each bead moved down on the top frame represents 'five'. The positions of the beads represent the numbers.

More than a hundred years ago Charles Babbage, an English mathematician, designed an analytical engine for performing calculations. It worked in decimal numbers but it was too complicated and cumbersome to build. ENIAC, the first electronic computer, also worked in decimal. Modern computers, however, all work in binary. Binary numbers need only two symbols. An electric current not flowing is represented by 0 and a current flowing is represented by 1. In this way a number is converted into a series of electric pulses. Each unit (0 or 1) is called a bit.

The modern computer has several parts. First there must be some way of getting information and instructions into the machine. This is the 'input'. A common method is to use a paper tape with groups of

Mathematical problem-solving illustrated in a 16th century German publication

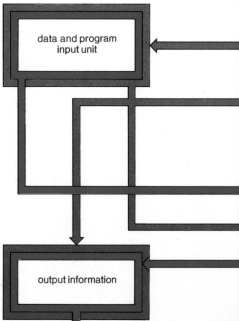

data and program input unit

output information

holes punched in it. Each group of holes represents a letter or number. A light shining through the holes in the moving tape produces the pulses of current. The numbers are then held in the store. There are various kinds of store. One is the magnetic core, made up of many small pieces of magnetic material and each piece stores one bit, according to which way the material is magnetized. Calculations are performed in the central processing unit (CPU) which contains electronic circuits for adding, subtracting, multiplying, dividing and other mathematical operations. The results are sent out of the computer through the 'output'. This may consist of a variety of devices, but often it is an electric typewriter that automatically types out the results onto a roll of paper.

The equipment and machines used for input, output and storage are called the hardware. The program, which consists of the instructions telling the CPU what to do with the input, is the software. Computers, which can calculate in minutes what would take a man years to complete, are used in many ways – for storing and sorting information, controlling machines and helping in scientific research.

computer layout

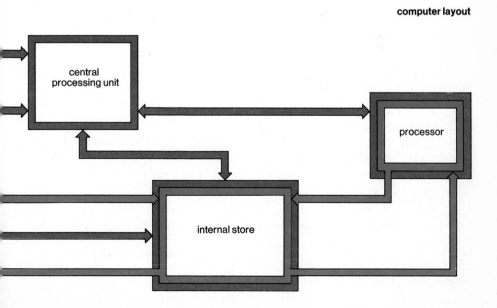

central
processing unit

processor

internal store

Many mathematicians are interested in investigating numbers for the sake of the numbers themselves. Usually, however, the numbers we use refer to numbers of things. Often they refer to measurements, such as metres, seconds or amperes.

To measure something there must first be something else to compare it with. The earliest measurements were based on the size of a man's body. The cubit was about the distance from the elbow to the tip of the middle finger. The yard was the distance from the nose to the tip of the outstretched arm. Since men's arms vary in length, accurate measurements have to be based on a permanent standard that is agreeable to everyone. Even today there is still confusion, with units of measurement meaning different things. The ton, for example, is 2240 pounds in Britain, but in the United States it is 2000 pounds. To make matters worse the metric ton (or tonne) is 1000 kg.

The first step towards a logical system was the metric system introduced by Napoleon in France in the late 1700s. The French wanted to use standards that would not change, so they defined the metre as one ten-millionth of the distance from the North Pole to the equator

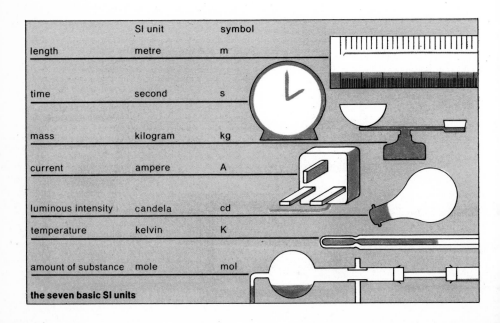

	SI unit	symbol
length	metre	m
time	second	s
mass	kilogram	kg
current	ampere	A
luminous intensity	candela	cd
temperature	kelvin	K
amount of substance	mole	mol

the seven basic SI units

along a meridian passing through Dunkirk, France. The gram was defined as the mass of one cubic centimetre of water at its freezing point. However, over the years these definitions proved unsatisfactory, and an international committee worked out a system of units to replace them. In 1960, by international agreement, SI units (Système International) were introduced and have since been adopted by most countries of the world.

The SI system has seven base units (including the metre, second, kilogram and ampere). All other units in scientific work are derived from these base units by simple definitions, and all measurements are now made in these units or in multiples of them. The base units themselves are precisely defined by scientific methods. Only the kilogram remains based on a standard block of metal (platinum-iridium) kept at Sèvres, near Paris. The metre, for example, is now defined in terms of the wavelength of a certain kind of light emitted by the gas krypton under certain conditions. The adoption of SI units throughout the world will make finance and commerce simpler when the difficulties of changing to it have been overcome.

In 1497 the Exchequer of Henry VII published a table which standardized weights and measures

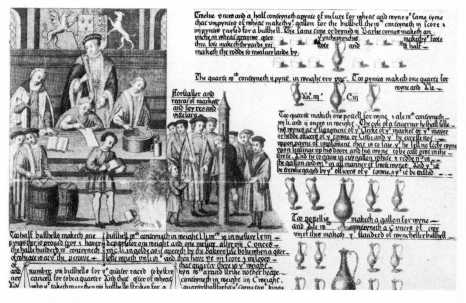

Mathematics
35   geometry and triangles

Geometry is one of the oldest branches of mathematics. Ancient Egyptian builders managed to build the Great Pyramid so that the length of its sides are accurate to within two centimetres because they had practical experience with the properties of lengths, angles and shapes. Vertical walls were built by using a plumb line and horizontal surfaces were laid with the help of a water level. The angle between the vertical direction and the horizontal is 90° – a right angle. They also knew a method of making right angles with three rods – one of three units in length, the other of four units and the third of five units. If these are put together they form a triangle in which the angle between the two shorter rods is 90°. It does not matter what units of measurement (such as metres or miles) are used as long as the sides are in the ratio of 3:4:5 – they will always produce a right-angled triangle.

Not all triangles contain right angles. One way of classifying triangles is by the lengths of their sides. A triangle whose three sides are equal is called an equilateral triangle. Triangles with two sides equal are called isosceles and those with three unequal sides are called

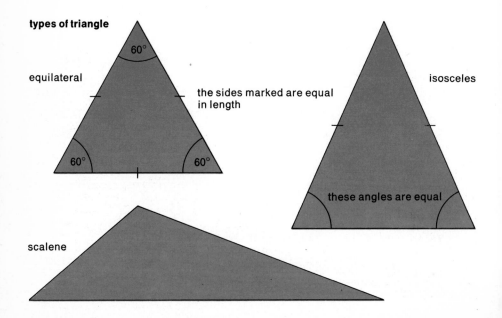

types of triangle

equilateral

60°

the sides marked are equal in length

60°   60°

isosceles

these angles are equal

scalene

scalene. Triangles are important in the construction of frameworks for roofs and aircraft, because they have the most rigid shape.

The ancient Greeks learned the beginnings of geometry from the Egyptians and then worked out a large number of characteristics of triangles and other shapes. In the third century B.C. Euclid, a professor at Alexandria, based his geometry on definitions of points, lines and angles, and also on certain hypotheses. These hypotheses were statements that were assumed to be true under all circumstances; for example, the statement that only one straight line can be drawn through two points. He then showed that other facts (theorems) followed from these hypotheses. The famous theorem of Pythagoras, a Greek mathematician, is a basic law that is true for any triangle containing a right angle. It states that the area of a square drawn on the side (hypotenuse) opposite the right angle is equal to the total area of the squares drawn on the other two sides. For a right-angled triangle with sides 3 and 4 units long, the hypotenuse must be 5 units long, because $4^2 + 3^2 = 5^2$. Other examples of right-angled triangles are those with sides 5, 12 and 13 units long and 7, 24 and 25 units long.

The Cinesphere in Toronto, Canada, is an example of a geodesic dome

Pythagoras' theorem

Mathematics
36   circles

Geometry is the study of shapes made up of straight lines and also the study of curves. The most familiar curve is the circle. It can be drawn with a compass or by sticking a drawing pin in paper, looping a string over it, attaching a pencil to the other end of the string and moving the pencil around the pin, keeping the string taut. It is obvious from the way that it is drawn that any point on the circle is the same distance from its centre. This distance is called the radius of the circle.

Ancient mathematicians thought that the circle was the perfect shape. A circle always has the same form even when it is rotated, which is why such things as wheels and gears are circular. Over two thousand years ago, Eratosthenes, a Greek mathematician, guessed that the Earth was spherical and found a way of measuring its circumference. He noticed that at midday at Cyrene the sun was directly overhead because its rays shone down a deep well. At the same time of day at Alexandria the rays were not vertical, but were inclined at an angle. By measuring this angle he was able to find that the angular distance between Alexandria and Cyrene was $7\cdot2°$. He knew the distance overland in modern units was 925 km and was able

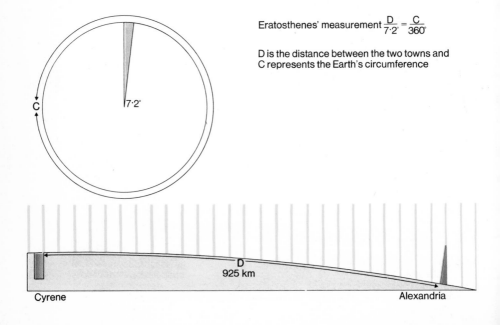

Eratosthenes' measurement $\dfrac{D}{7\cdot2°} = \dfrac{C}{360°}$

D is the distance between the two towns and C represents the Earth's circumference

$7\cdot2°$

C

D
925 km

Cyrene

Alexandria

to work out a value of 46 250 000 metres for the Earth's circumference. The accurate value is 40 089 350 metres.

Circles were widely studied in ancient times because of their properties. Mathematicians in Egypt, Babylonia and China knew that the distance around any circle (circumference) is roughly three times its diameter (a straight line passing through the centre of the circle and joining two points on the circumference). In every circle, no matter how small or how large, its circumference divided by the diameter is always the same. This ratio of circumference to diameter is represented by the symbol π, the Greek letter pi. Archimedes found that the value of π was slightly greater than 3 – it is about $3\frac{1}{7}$ – and we often use this figure for rough calculations. For more precise work the value is 3·14159 . . . where the decimals go on forever without repeating. Pi is what is known as an irrational number. We do not know its value exactly but can calculate it to any required accuracy – an early computer worked it out to 10 000 decimal places.

The circumference of a circle is $\pi \times$ its diameter, or $2\pi r$ where r is the length of the radius. The area of a circle is therefore πr^2.

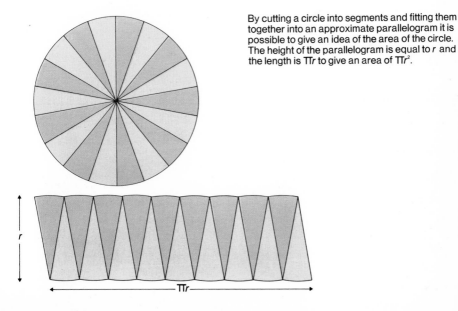

By cutting a circle into segments and fitting them together into an approximate parallelogram it is possible to give an idea of the area of the circle. The height of the parallelogram is equal to r and the length is πr to give an area of πr^2.

Because the Greeks had a mystical idea of the perfection of the circle, it was thought for centuries that the planets moved in circles in the sky. We now know that planets move around the sun in another type of curve, called an ellipse. An ellipse can be drawn by putting two drawing pins in a piece of paper and passing a loop of string around them. The point of a pencil is placed in the loop and moved around the pins with the string held taut.

Each of the points at which the pins are placed is called a focus. If the distance between the pins is changed, the shape of the ellipse changes. As the foci get closer together the ellipse becomes more like a circle. Eventually, when the two foci occupy the same point a true circle is formed.

If a circular disc is held at an angle, the shape seen is an ellipse. An ellipse is also produced if a beam of light is shone on a wall at an angle.

Another way of producing an ellipse is by cutting a cone. A cone has the shape of an ice-cream cornet: it has a circular base and the sides taper to a point. In a right circular cone this point (vertex) is directly above the centre of the base. If a right circular cone is cut

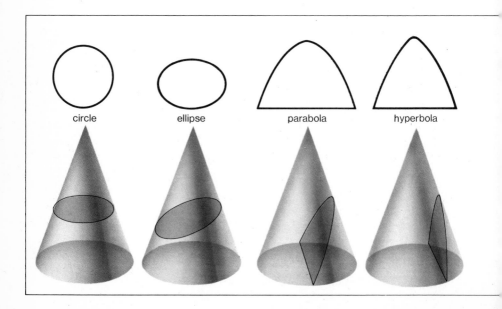

circle   ellipse   parabola   hyperbola

across horizontally a circle is formed. If it is cut at an angle an ellipse is produced. For this reason circles and ellipses are called conic sections. It is possible to cut a cone in such a way that a closed curve is not produced. If it is cut vertically, the curve produced is a hyperbola. A parabola is formed when a cone is cut along a slanting plane that passes through its base. These two curves are also conic sections. Hyperbolas can have two parts. This is because a cone can consist of two ice-cream cornets which meet at their sharp points – so the whole figure has the shape of an X. If such a cone is cut vertically, two hyperbolic curves are formed.

When a cricket ball, or any other projectile, moves through the air it travels along a parabolic path. A parabola rotated about an axis sweeps out a surface in space called a paraboloid. Telescope mirrors and the dishes of radio telescopes are made in this shape. They can reflect a parallel beam of light or radio waves to a single point which is the focus of the parabola. Parabolic shapes are also used to produce a parallel beam from a small source, as in searchlight reflectors, car headlamps and electric heaters.

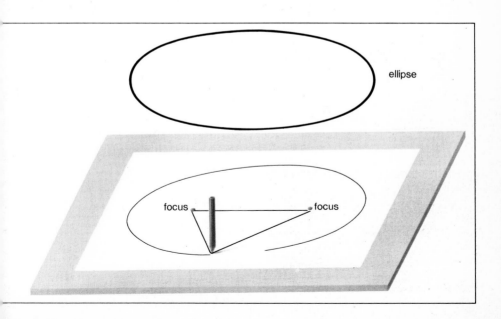
ellipse
focus   focus

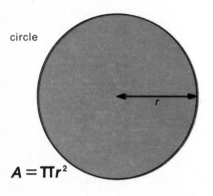

triangle

$A = \frac{1}{2}bh$

circle

$A = \pi r^2$

parallelogram

$A = lh$

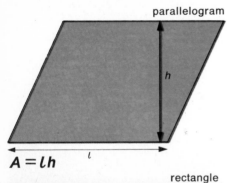

rectangle

$A = lb$

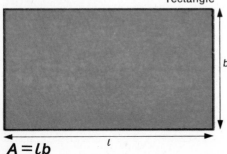

areas of simple figures

If a car travels for two hours at 60 miles per hour it covers 120 miles. Similarly, a car moving at 50 miles per hour for three hours travels 150 miles. The distance is obtained by multiplying the velocity by the time.

This statement can be written in a much shorter way if symbols are used instead of words. If distance is represented by the letter s, velocity by the letter v, and time by t, then we can write $s = v \times t$. We usually do not bother to include the multiplication sign and simply write $s = vt$. This type of expression belongs to algebra, a branch of mathematics which was invented by the Arabs. The word comes from the Arabic 'Al jebr' meaning the great art.

Algebra is a form of language for writing about numbers and measurements and the relationships between them. Not only is it much shorter than using words but it helps in working out mathematical problems. For example, in the equation $s = vt$, everything on the left of the equals sign is equal to everything on the right. We can do anything to one side as long as we do the same to the other. For example, if both sides are divided by v we get

$\frac{s}{v} = \frac{vt}{v}$ or $t = \frac{s}{v}$; i.e. the time taken is the distance divided by the velocity.

A more complicated example would be a problem such as 'in two years' time a man will be twice as old as his son is now. If their total age is 67, how old is the son?'

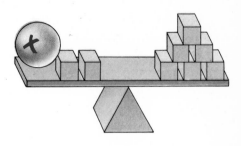

X + 2 = 6

First we translate the words into symbols. If m stands for the man's age in years, then in two years' time he will be $m + 2$ years old. Now we can construct our first equation: $m + 2 = 2s$, where s is the son's age. We now have one equation, but two unknown symbols, i.e. m and s. In order to solve the equation, that is to assign values to the symbols, we must always have the same number of equations as we have unknowns. In this case, therefore, we need one more equation: it is $m + s = 67$. If $m + 2 = 2s$ we can subtract 2 from each side and get $m = 2s - 2$. We can now use this value of m in the equation $m + s = 67$, thus $2s - 2 + s = 67$. Adding up the $2s + s$ on the left-hand side and putting the numbers on the right we get $3s = 69$, or $s = 23$. The son's age, therefore, is 23.

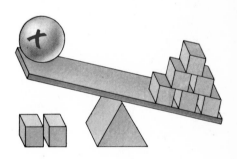

X is smaller than 6

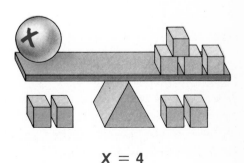

X = 4

Mathematics
39 co-ordinate geometry

In 1637 a French mathematician, René Descartes, invented co-ordinate geometry, a method of combining algebra and geometry which uses graphs to express the relationship between two quantities. First two lines are drawn at right angles to each other. The horizontal line is called the x-axis and the vertical line is called the y-axis. The point at which the lines cross is called the origin, O, and all distances are measured from this point. The two axes are marked with scales, so that distances upwards and to the right of the origin represent positive values of x and y. Distances downwards and to the left of the origin represent negative values. If we start at the origin and move two units right along the x-axis and three units upwards parallel to the y-axis, we reach a point which is labelled (2,3). These numbers are the co-ordinates of the point: the x-value is always written first. The point (2,3) is different from (3,2).

Because this form of mathematics relies on co-ordinates, it is called co-ordinate geometry, and it can be used to represent algebraic equations. The simple equation $y = 3x$, means that when $x = 1$, $y = 3$; when $x = 2$, $y = 6$. If we mark the points (1,3) (2,6) between the axes they can be connected by a straight line. Every point on the line is a point at which $y = 3x$. Graphs can illustrate the connection between two quantities. For example a graph is often used for changing a temperature in degrees Celsius (C) into degrees Fahrenheit (F), based on the equation $F = 1 \cdot 8C + 32$.

The equation determines the type of line produced. In the illustration a graph of the line $y = 2x$ is drawn; it will not slope as much as the line $y = 3x$. The gradient of a line is the distance it rises divided by the distance travelled horizontally: $y = 2x$ has a gradient of 2, $y = 3x$ has a gradient of 3. The graph of the line $y = 3x + 2$ does not pass through the origin. Instead it cuts the y-axis at the point (0,2). In general, all straight lines have an equation of the form $y = mx + c$, where m is the gradient and c is the point at which the line cuts the y-axis. Other more complicated equations lead to different curves. For example $x^2 + y^2 = r^2$ is a circle of radius r with its centre at the origin. It is also possible to use co-ordinate geometry in three dimensions; in this case a z-axis is added to the x- and y-axes.

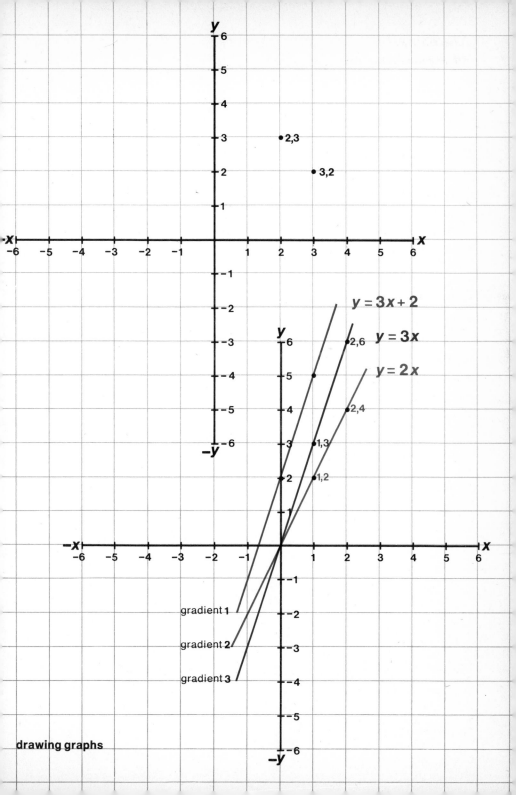

drawing graphs

It is easy to find the area of a rectangle. Simply multiply its length by its breadth. However, for a figure with curved sides the method of finding the area is not as easy.

The curve in the diagram is part of a parabola with an equation $y^2 = 4x$. One way of finding its area is to draw vertical lines at the points $x = 1$, $x = 2$, $x = 3$ and $x = 4$. This divides the area under the curve into five parts. Now we can construct four rectangles, as in the second diagram. Each rectangle is one unit wide and its height can be found by co-ordinate geometry using the curve's equation. For example, when $x = 1$, $y^2 = 4 \times 1$, and therefore $y = \sqrt{4}$. Thus the area of the smallest rectangle is $1 \times \sqrt{4}$. The areas of the other rectangles can be found in the same way and all four added together to give a total area of $(1 \times \sqrt{4}) + (1 \times \sqrt{8}) + (1 \times \sqrt{12}) + (1 \times \sqrt{16})$.

Of course, this area is not exactly the area we require. It is smaller than the true area by the amount coloured tan in the diagram. If we were to divide the x-axis into nine narrower rectangles (third diagram) we would get a closer answer and there would be less tan in the diagram. In fact, the more narrow rectangles we take the nearer

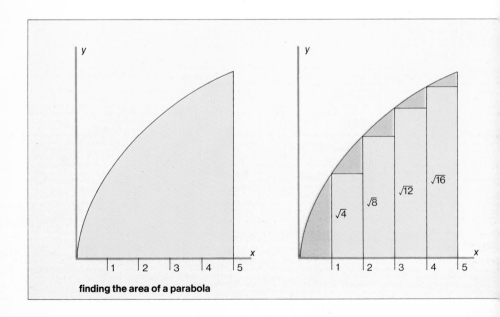

finding the area of a parabola

the sum of their areas approaches the area under the parabola.

The total area of the rectangles can never be greater than the area under the curve, but it gets closer and closer to it as we take more and more strips. We say that it tends to a limit at which it equals the parabola's area. If you can imagine adding an infinite number of strips of zero width then this is the limit, at which the approximate area exactly equals the true area.

In fact it is possible to work out a formula for the area by using the curve's equation and this idea of limits. Then the area can be calculated from the equation of the curve without having to draw a diagram and make actual measurements.

The branch of mathematics in which this method is used is calculus. It was invented in the 17th century independently by the English and German mathematicians Isaac Newton and Gottfried Leibniz. This branch of calculus, concerned with measuring areas and volumes, is integral calculus as it depends on the process of integration, or adding together a large number of small elements. The other branch, differential calculus, is concerned with rates and speeds.

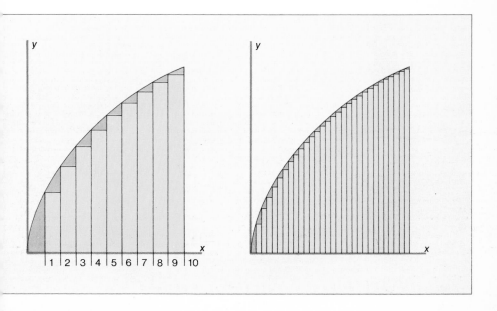

Ordinary algebra is a form of language for saying things about numbers and the way they are combined. Boolean algebra is a different type of algebra for describing collections of things. It was invented by a 19th-century English mathematician, George Boole.

Suppose that in a group of 25 students, 17 study mathematics and 16 study science. Obviously some must be studying both. In fact 8 students $(17+16-25)$ study both mathematics and science. In Boolean algebra the full group of students is called the universe set, P. Those studying mathematics are a sub-set, which can be given the symbol M; in the same way S is the sub-set of science students.

Boolean algebra has its own system of signs and symbols. Just as numbers are combined by addition, multiplication, etc., so sets are combined in certain ways. The full set of students P consists of a number of individuals P_1, P_2, P_3, etc. This is written $(P_1, P_2, P_3,$ etc.$) = P$. In this case $P_1 \in M$, etc., meaning that P_1 is a member of the set M. If we write $M \subset P$, we mean that the sub-set M is contained in the universe set P. The intersection of the sub-sets M and S is written $M \cap S$ (read M cap S). It is the set of students who study

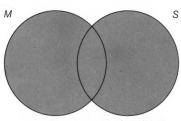

the intersection of M and S ($M \cap S$)

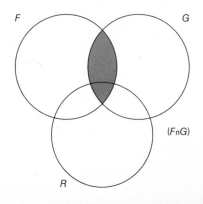

the union of M and S ($M \cup S$)

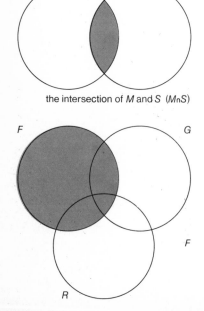

both mathematics and science. Another example is the intersection of the set of all schoolteachers with the set of all people with red hair: it is the set of all red-haired schoolteachers.

The total set of all students is given by the union of M and S. This is written $M \cup S$ (read M cup S). It is the set of students who are studying either mathematics or science or both. The operations of cap and cup are most easily understood by using Venn diagrams in which a set of things is represented by the area within a circle, and by the areas of the overlapping parts of overlapping circles.

The operations of intersection and union are rather like multiplication and addition in arithmetic. If F, G and R are three sets it is possible to show that $F \cap (G \cup R)$ is the same thing as $(F \cap G) \cup (F \cap R)$. You can see this from the Venn diagram. This is like the law that $3 \times (2+7) = (3 \times 2)+(3 \times 7)$. Another basic rule of Boolean algebra is that $F \cup (G \cap R) = (F \cup G) \cap (F \cup R)$. Note that in arithmetic $3+(2 \times 7)$ is not the same as $(3+2) \times (3+7)$.

Boolean algebra can also be applied to statements in logic as well as sets. Its modern application is in computer programming.

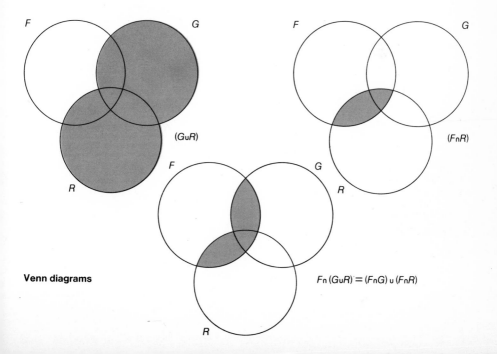

Venn diagrams

Mathematics
42   topology: the geometry of position

Euclid's geometry was mostly concerned with size and shape. Topology is the branch of geometry that is only concerned with the study of positions and boundaries. In a fair-ground distorting mirror, for instance, your features do not change their relative positions, and your two eyes stay on each side of your nose. Changes in shape are called deformations or transformations and topology is the study of properties that do not change with deformation.

A famous topological problem is the puzzle of the seven bridges in the Russian city of Königsberg. Two islands in the River Pregel are joined to each other, and to the two river banks, by seven bridges. The problem was to plan a walk in which each bridge was crossed only once. Leonhard Euler, a Swiss mathematician who was invited to map out a route, showed that it was impossible. Euler simplified the map into a network. The four points represent the river banks and the islands. The seven lines, or branches, are the seven bridges. There must be an even number of branches to each point, so that the arrivals balance the departures: otherwise one branch would have to be crossed twice over. The only exception to this rule is at the

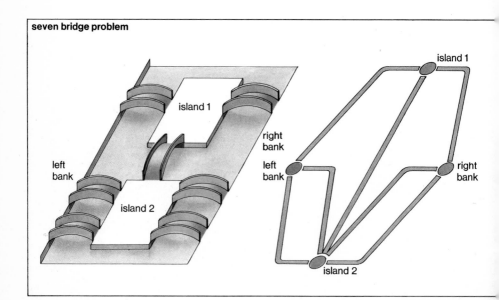

seven bridge problem

island 1

island 1

right
bank

left
bank

left
bank

right
bank

island 2

island 2

beginning and end of the journey. In other words, in order to trace out the network in a single trip, all but two points must have an even number of branches. This rule can be tried out on other networks.

In topology, surfaces are classified into genuses. A surface of genus O is one in which a closed curve breaks the continuity of the surface. Imagine the surface of an apple with a circle drawn around it. A spider climbing down the stalk could not reach the bottom of the apple without crossing the circle. A map in which countries with a common frontier are coloured differently can be drawn on a genus O surface using only four colours, no matter how many different countries there are.

A surface of genus 1, however, can have one closed curve without breaking the continuity between any two points. Imagine the apple with the core removed and a length of cotton threaded through the hole. If the cotton is tied tightly over the surface to form a closed curve, a spider at one point on the surface of the apple could now reach any other point without crossing the cotton. For genus 1 surfaces the maximum number of colours required to draw a map is seven.

four colour problem

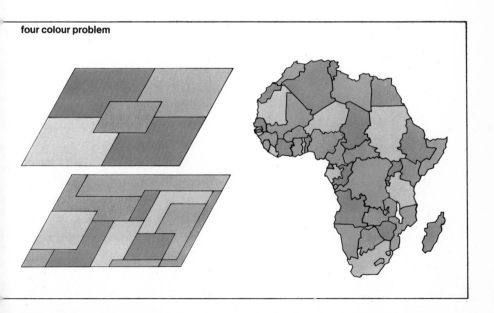

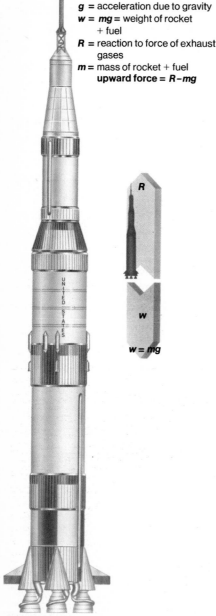

g = acceleration due to gravity
w = mg = weight of rocket
+ fuel
R = reaction to force of exhaust
gases
m = mass of rocket + fuel
upward force = $R - mg$

R

w

$w = mg$

Saturn 5 rocket showing that the upward thrust is produced by the reaction to the exhaust gases. As the fuel is used up, w is reduced and R increases.

Mechanics is the branch of science that deals with forces and their effects. Forces are simply pulls or pushes. Imagine two teams in a tug-of-war exactly balanced in strength. They are said to be in equilibrium. One man slips and the other team pulls away to win because when one force – the strength of the one man – is removed, the equilibrium is upset. A body at rest, in equilibrium, moves only when some force is removed or a new force added. A billiard ball motionless on a table will move when some new force is applied, for example, when another ball strikes it. Isaac Newton expressed this first law of motion as follows: if a body is at rest it will remain at rest, or if it is moving in a straight line at constant speed it will continue to do so, unless a force acts on it.

In his second law of motion, Newton showed that a force is mathematically equal to the acceleration it produces in a body multiplied by the mass of the body. His third law says that for every force there is an equal force in the opposite direction. The opposite force is called the reaction. If you put a book on a table, its weight is a downward force on the tabletop. The reaction is the upward force that keeps the book in equilibrium. If there was no reaction the book would have to move downwards.

Everything is attracted to the Earth by the force of gravity. What we call weight is the measure of the force of gravity on an object. If you weigh an object in a valley and again at the top of a mountain, its weight will differ because the pull of gravity diminishes as the distance from the Earth's centre increases. Mass is not the same as weight although in everyday use they are often given the same meaning. The mass of a body is a measure of the amount of matter it contains; weight is a force. This may, at first, seem strange but an astronaut in space quickly realizes it. His body still has the same mass as it does on Earth, but the pull of gravity is so remote that he feels absolutely weightless and objects in the spacecraft float around him.

Mass and weight are measured in different units in science, although for everyday purposes the same units are used. Mass is measured in kilograms (in SI units), whereas weight is measured in the units of force – newtons.

When an astronaut takes a walk in space he feels weightless

If, in a tug-of-war, the opposing teams are perfectly balanced neither side will move because they are in equilibrium. If one man drops out, the stronger side will easily be able to pull away in a straight line. But what happens if the forces are not acting in exactly opposite directions, but at an angle to each other, as for example when two boys are pushing at a large beachball? If the forces do not balance, the object must move, but it cannot move in two directions at once. It therefore moves in the single direction that results from the combined action of the two forces – it moves in the direction of the resultant force.

Exactly the same problem arises when an aircraft is flying in a crosswind. Two forces are involved : there is the force of the engine driving the plane forwards and another force, due to the wind, forcing the aircraft sideways. The navigator must make allowance for the effect of the wind force or the plane will be blown off its course. In the diagram the plane sets out to fly to a destination to the north-east. The wind is blowing from the south. The navigator draws a parallelogram in which the direction and speed of the wind (its velocity) produce a resultant in the required direction. He draws a parallelogram

wind

drift angle

of velocities (velocity is speed which has a given direction). Until quite recently the diagrams had to be drawn by the navigator by hand, but in modern aircraft the information is supplied to a computer and a chart drawn up so that the course can be set automatically.

If you want to row across a fast-flowing river you will have to work out a similar navigation problem in order to land at a particular place on the opposite bank. If you want to land exactly opposite, the boat must be pointed upstream. The faster the river is flowing the further upstream it will have to be pointed.

The exact resultant of any two forces can be found easily by using the parallelogram of forces, which is similar to the parallelogram of velocities. Draw two sides of a parallelogram to represent in direction and length the directions and sizes of the two forces. Complete the parallelogram and the diagonal gives the direction and size of the resultant force. A force or velocity which has a known direction is called a vector, so the parallelogram of forces or the parallelogram of velocities is sometimes called a vector diagram. (A quantity that has only a stated magnitude but no stated direction is called a scalar.)

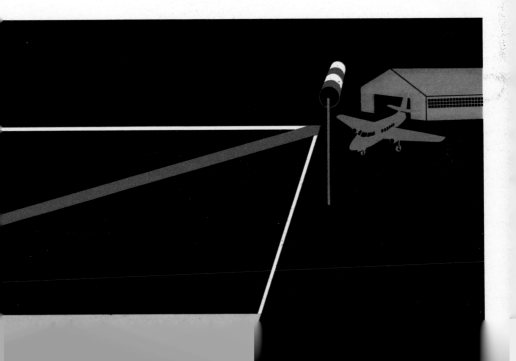

Mechanics
45 *work, energy and power*

Work – physical work – is moving something from one place to another. This involves overcoming some force or resistance, such as the force of gravity when going upstairs or the resistance of water in swimming. In scientific terms work is defined as a force multiplied by the distance through which it moves. If we carry a suitcase up two equal flights of stairs we do twice as much work as carrying it up one of the flights. To provide the force to do this requires energy: when we lift things the energy comes from the food we eat which enables our muscles to do the work. In the world around us most work is done by machines whose energy comes from fossil fuels, such as coal or oil. Other sources of energy are the wind, moving water in rivers, the rise and fall of tides and, most important of all, nuclear energy.

Energy is of two kinds: stored (potential energy) or energy of movement (kinetic energy). Potential energy is the energy of position. A mass m when it is raised to a height h has a potential energy of mgh, where g is the acceleration due to gravity (acceleration of free fall). When you wind up a watch you store energy in the spring. When this potential energy is released it becomes kinetic energy as the clock-

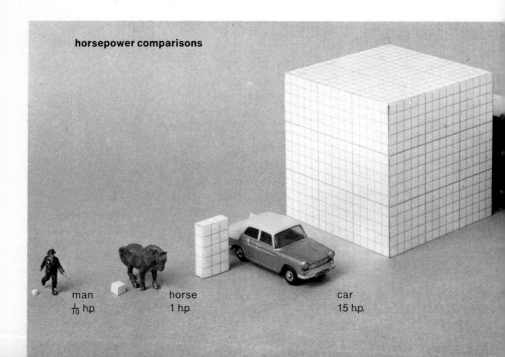

horsepower comparisons

man
$\frac{1}{10}$ h.p.

horse
1 h.p.

car
15 h.p.

work turns. The explosive in a cartridge contains potential energy and when it is fired it shoots the bullet at high speed from the gun. The bullet has kinetic energy. The kinetic energy of a moving body depends on the mass of the body and its velocity. The formula for kinetic energy is $\frac{1}{2}mv^2$ (where m = mass and v = velocity).

The same amount of work can be done either slowly or quickly. You can drive a car one kilometre in a minute on a motorway or one kilometre in two minutes in town. The amount of work the engine does (moving the car one kilometre) is the same in both cases. But to do it more quickly calls for more power (and you will use more fuel at 60 km per hour than at 30 km per hour).

The Scottish inventor, James Watt, decided to establish some standard for measuring power. He found that a horse could raise a 1000 lb weight 33 ft in a minute and so one h.p. was defined as 33 000 foot-pounds per minute. In his honour, when a measure for electrical power was needed it was called a watt. One kilowatt (1000 watts) is equal to 1·34 h.p. The watt is now used to measure all types of power because it is based on metric units and horsepower is not.

locomotive
3500 h.p.

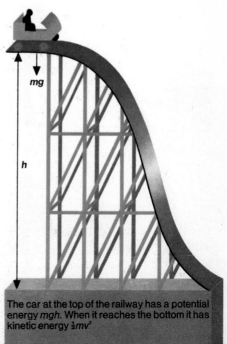

The car at the top of the railway has a potential energy mgh. When it reaches the bottom it has kinetic energy $\frac{1}{2}mv^2$

Mechanics
46   levers

Machines are usually thought of as rather complicated and intricate devices, but to an engineer a machine is any arrangement that enables resistance to be overcome at one point by applying a force at some other point. Some machines are very simple: for example, a lever, such as a stick or a metal bar, is a machine that helps a man to raise a load which he would be unable to move by using only his unassisted hands and muscles. Neither a lever, nor any other kind of machine, increases a man's strength or an engine's power: their purpose is to redirect the effort to where and when it is needed.

Levers can be arranged in various ways, but in the most common form the pivot, or fulcrum, comes between the load and the effort. To obtain what is called a mechanical advantage, the distance between the load and the pivot must be shorter than the distance between the effort and the fulcrum. If the load is 30 cm from the pivot and the effort is 90 cm from the pivot, the mechanical advantage is $\frac{90}{30} = 3$. But the load will move only one-third of the distance that the effort moves. By lengthening the handle of the lever the mechanical advantage can be further increased, but the effort has to move a

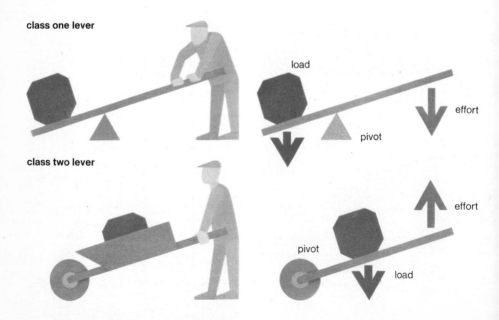

class one lever

load

effort

pivot

class two lever

effort

pivot

load

greater distance. If you cannot remove the lid of a paint tin with a coin as a lever (giving you a mechanical advantage of perhaps four) you can take a screwdriver, which will give you an advantage of 20 or 30, and the lid comes off quickly.

A screwdriver by itself is not a machine, but a tool. It becomes a lever only when it is placed on a fulcrum, such as the side of the paint tin, so that one part is longer than the other. If the two arms of a lever are equal no mechanical advantage is gained.

There are, in fact, three classes of levers. In a class one lever, the fulcrum is between the load and the effort. In a class two lever – in a wheelbarrow for example – the load comes between the fulcrum and the effort. In a class three lever, the effort comes between the load and fulcrum – the forearm pivoting about the elbow is an example.

The bones of the body act as levers to produce movement supplied by the contraction of muscles. Muscles can only pull when they contract: they cannot push. So there must be two muscles to control each movement; one to pull the lever one way and one to pull in the opposite direction.

A device for raising water from a well and the oar are both examples of simple levers

class three lever

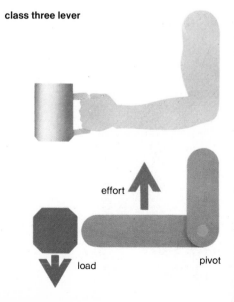

effort

load

pivot

Mechanics
47   wedges, inclined planes and screws

A ramp and a zigzag mountain path have something in common with
a chisel and an axe; they all use the principle of the slope, or inclined
plane, to obtain a mechanical advantage. The ramp and the climbing
road are stationary and act as machines by making it easier to raise
a load from a lower level to a higher one. Chisels and axes, like all
cutting edges, are wedge-shaped and only act as machines when they
are moving. They are all machines because they make it possible for
a small effort applied over a large distance to raise a large load by a
small amount.

If an inclined plane is coiled around into a spiral, like a winding
staircase, it becomes a screw. This is very useful because a long
inclined plane gives a much greater mechanical advantage than a
short steep one. The mechanical advantage of any particular screw
depends on its pitch – the distance between the crests of two adjacent
threads. This is the distance that a screw will move forward when it is
given one full rotation.

There are two main ways of using the screw principle: for fixing and
holding things together and for raising a load. Ordinary wood screws,

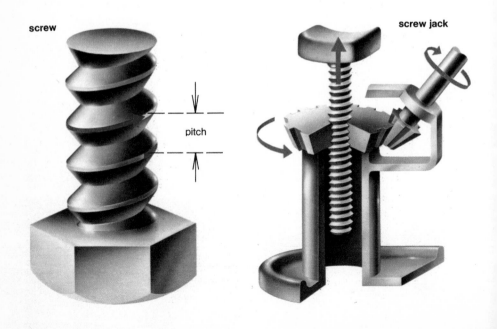

screw

pitch

screw jack

a carpenter's vice, a press and the screw cap of a bottle use the first principle. A screw jack for lifting a car, a ship's propeller and a grinding machine are all in the second category. In a ship's propeller the tip of each blade traces out a curve called a helix as the ship moves forwards. The pitch of the propeller is the forward distance through which each blade moves in one complete revolution. If water was a completely solid and unyielding substance, the ship would move forwards by a distance equal to the pitch for each revolution, but in water it only moves about 60 or 70% of this distance.

A screw can also be used to make a very accurate measuring instrument, such as an engineer's micrometer screw gauge. If a screw has a pitch of one millimetre it has to twist one full turn to move one millimetre forwards. If it is twisted only half a turn it moves forwards only half that distance, and so for each fraction of a turn it advances the same fraction of a millimetre. This provides engineers with a way of making very accurate measurements. The marks that show how much the barrel of the micrometer has been turned also show the exact thickness of the object being measured between its jaws.

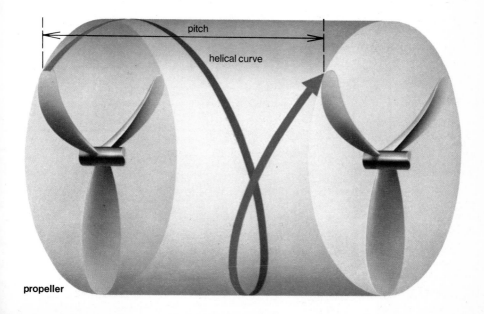

propeller

Levers, wedges and inclined planes were in use long before the Greeks. Years later the Sumerians and Egyptians used wheels, but it took many hundreds of years before rough, solid wooden rollers were developed into wheels that could revolve freely on their axles.

A pulley is simply a wheel with a rope running around it. Slinging a rope over the rim of a wheel makes it easier to raise a load. With suitable arrangements of pulleys and ropes a man cannot only lift several times his own weight but can also hoist the load to many times his own height.

A single pulley lets the rope run easily and changes the direction of the pull, but it does not provide a mechanical advantage. Two pulleys together, however, double the force applied to the load so that the same effort lifts a double weight. But no energy is saved because the rope to which the effort is applied has to be pulled twice the distance that the load is raised. Indeed, some energy is wasted in friction.

Although they seem to be very different, pulleys and levers work on exactly the same principle. They both turn on a pivot, the lever on its fixed fulcrum and the pulley on its fixed centre. A pulley is a lever with its fulcrum in the middle and its two arms on either side, like two spokes making a diameter. The advantage of a pulley is that, unlike a lever, it can turn continuously.

The pulley system used by a mechanic to remove the engine from a car is a block and tackle. With three fixed pulleys and two movable ones a pull of only one-fifth the weight of the load is needed. But five metres of rope must be pulled for every metre the load is raised.

A ship's capstan, the winch of a crane and the pedal crank of a bicycle all combine levers and wheels and provide a turning effect. Exactly how much turning power they have depends on how far away the force is from the turning point. Engineers call the turning power of a force the moment of that force. The moment of a force is equal to the product of the force and the perpendicular distance between the line of action of the force and the axis of rotation. So the moment of the force of your foot on a bicycle pedal is equal to the force of your foot multiplied by the length of the pedal crank.

An early engraving shows a system of pulleys for raising loads

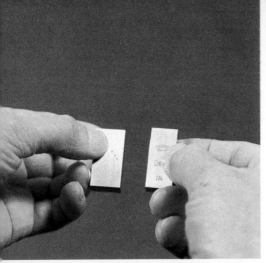

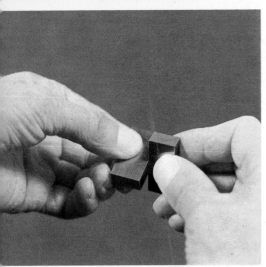

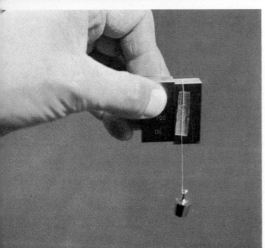

When what seems to be a smoothly-polished surface is examined under a microscope it is seen to be anything but flat. It looks as if it is made up of hills and valleys. If two such surfaces are put together, they touch at only a few places. When one of the surfaces is dragged over the other, the hills and valleys lock with each other and so resist the movement. The harder the two surfaces are pressed together the more firmly the hills and valleys grip, and the greater the resistance. This resisting force is called friction. It is friction that brings a car to a stop when the brake shoes press against the drums, and which makes a car move when the clutch plates engage.

All surfaces that come into contact with each other, even a bullet whizzing through the air or a ship gliding through water, experience friction. They are not only slowed down but some of the energy of motion is converted into heat. The heat generated by friction when a meteorite enters the atmosphere is almost always enough to burn it away before it reaches the ground. Designers of

Friction between two smooth non-magnetic metal blocks can support a weight of 5 grams

cars, ships and aircraft try to cut down friction losses from air and water resistance by streamlining; the shape of a fish or a bird shows that streamlining is much older than an engineer's design.

Friction in machines is also destructive and wasteful: it causes the moving parts to wear and it produces heat where it is not wanted. Engineers reduce friction by using very highly-polished materials and by lubricating their surfaces with oil or grease. They also use ball bearings and roller bearings because rolling objects cause less friction than sliding ones. Although friction is very often an expensive nuisance, it would be a very strange world without it. With no friction, walking would be impossible because if we once started to move, we would not be able to stop. Nails and screws would never hold, a painter's ladder would slide to the ground and scaffolding would collapse. We also use the heat produced by friction to start a tiny explosion when we strike a light. Primitive man rubbed two sticks together until the friction made them hot.

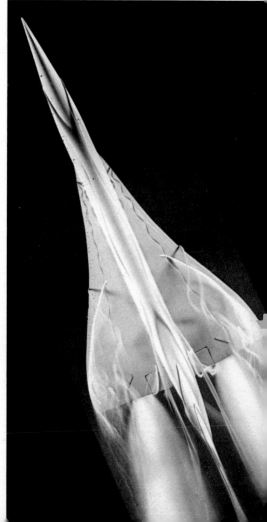

Top: *brown booby, tailored for flight.* Below: *streamline flow around supersonic aircraft*

If you want to move an object that is some distance away, it can either be pulled with a rope or pushed with a rod. These are simple ways of transmitting power. In most machines the power produced by the engine has to be transmitted to some other point in much the same way, by belts, chains, shafts or rods. In a motor car the engine turns a series of shafts which transmit the power to the wheels. A fan belt transmits engine power to the fan which cools the radiator and usually it also turns the dynamo or alternator. It is impractical to transmit power by such means over anything but very short distances because long belts or shafts are very cumbersome. Sometimes power can be transmitted over longer distances by pumping air or water through pipes – this is called a pneumatic or a hydraulic system.

For long distances electric energy is the most suitable method of transmitting power from where it is generated to where it is needed. In many cases an engine or turbine is used to turn a generator, which converts the mechanical energy into electrical energy. The current produced by the generator is then carried by electrical cables to some distant point, where it can be used to operate an electric motor which converts it back to mechanical energy.

Almost all engines deliver power in the form of rotation. To produce a change in the speed of a rotating shaft or wheel, gear wheels can be used. Suppose that on the chain wheel turned by the pedal on a bicycle there are 48 teeth and on the chain wheel of the rear wheel there are 16 teeth. Every time the pedals are made to go around once, the back wheel will revolve three times. A gear with many teeth driving one with fewer teeth produces an increase in speed, while the opposite arrangement produces a reduction in speed. Gear boxes on cars are complicated devices for bringing gears of different sizes in contact with the engine shaft to produce changes in the speed of rotation of the wheels relative to the engine speed. When a gear or system of belts or chains increases the speed of a shaft, this does not mean that the shaft can do more work, in fact it can do less, because more energy is lost in friction. The principal benefits of these devices are that any desired speed of rotation can be arranged while the engine continues to rotate at its most efficient (economical) speed.

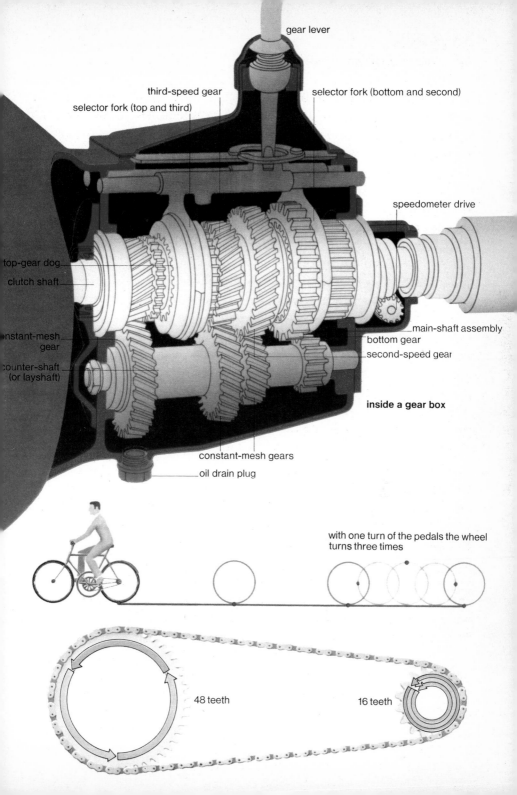

gear lever

third-speed gear

selector fork (top and third)

selector fork (bottom and second)

speedometer drive

top-gear dog

clutch shaft

constant-mesh gear

counter-shaft (or layshaft)

main-shaft assembly

bottom gear

second-speed gear

inside a gear box

constant-mesh gears

oil drain plug

with one turn of the pedals the wheel turns three times

48 teeth

16 teeth

For thousands of years primitive men had only their own muscles as a source of power. Later they domesticated the donkey and ox to help them. Later still they discovered that two natural sources of power – wind and flowing water – could also be put to work. Water can sometimes be controlled by canals and dams. The wind cannot be controlled but when it blows it can be used by sailing ships and windmills. Windmills were used for grinding corn and pumping water. To make use of the power from the slow-turning sails of windmills, gear wheels and other mechanisms had to be devised to vary the speed and direction of the revolving wheels.

Flowing water can be used as a source of power by placing a large wheel in the stream. Fins or buckets attached to the rim of the wheel make the wheel rotate as the stream of water moves.

The Romans invented several types of water-wheels. In some, the water fell from above (overshot wheels), in others a fast stream pushed at the bottom (undershot wheels). Gradually more and more efficient types were made, and finally came the water turbine, which makes the best possible use of the energy of moving water. Fast-

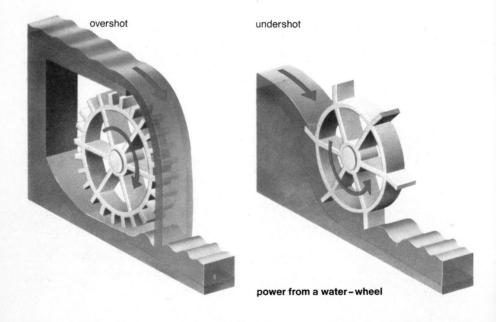

overshot   undershot

power from a water–wheel

flowing streams can supply enough energy to turn a water-wheel operating a mill or machine. Large waterfalls, where thousands of tons of water fall a great distance, can provide enough power to supply a whole town. Today water turbines are used in the great hydro-electric power stations that generate electricity in countries such as the United States and Switzerland. At Niagara Falls, in the United States, some 6 000 000 kg of water fall about 60 metres every second, sufficient to generate 3500 megawatts (million watts) of power, or about 4 500 000 horsepower. In other countries dams are built to ensure a constant supply of water which is channelled through huge pipes to a lower level where they turn turbines which then rotate generators. The Kariba Dam in Africa holds back the great Zambezi River to form a vast lake 5000 sq km in area. Another form of water power is provided by the rise and fall of tides, but this is very difficult to use, and ways of doing it are only now being developed. The first tidal power station has been built on the Rance estuary in France. These methods of obtaining energy from wind and water are being used again as fossil fuels become more expensive.

The windmill and water-wheel are sources of energy

Rods and belts by themselves can only transmit pushes and pulls in one direction: the great advantage of using liquids is that they transmit power equally in all directions at once. If you fill a rubber ball with water, prick it with holes of approximately equal diameter and squeeze it, water will spurt out in all directions with the same force.

The braking system of a car is a good example of how a hydraulic system works. When the brake pedal is pressed a piston operates which forces brake fluid out of the master cylinder and along four narrow pipes to the slave cylinders attached to the brake drums or discs so that the same pressure is applied to the brakes in each wheel. This brings the car to a smooth halt. Provided the system is kept filled with brake fluid, hydraulic brakes work instantly because liquids cannot be compressed to any great extent.

If air leaks into the system, the brakes become much less efficient. This is because, unlike liquids, gases are compressible and some of the movement of the brake pedal is taken up in squeezing the air bubble.

However, the more a gas is compressed, the greater is its resistance to further compression. For this reason compressed air can be used

car braking system

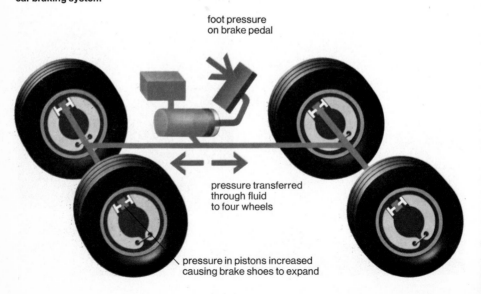

foot pressure
on brake pedal

pressure transferred
through fluid
to four wheels

pressure in pistons increased
causing brake shoes to expand

to transmit power in place of a liquid. In fact, some brakes are worked by compressed air while others, on electric trains for example, are vacuum operated.

The hydraulic ram, or press, is an example of how a small weight can produce a large force. If a mass of 10 kg is placed on a small piston in a tube of one sq cm area that is filled with fluid, the pressure throughout the fluid will be 10 kg per sq cm (or more correctly 98 newtons per sq cm). If this narrow tube is connected to a wider one of 50 sq cm area, the pressure of 10 kg per sq cm acting on an area of 50 sq cm will produce a force of 500 kg (4900 newtons). In this way a load just less than 500 kg can be raised, but only a tiny distance. If the 10 kg weight falls 10 cm, the 500 kg will only rise one-fifth centimetre.

This is what happens in the operation of a tip truck; a series of small pushes from a pump builds up a high enough pressure to raise the back of the lorry. The same principle works in the hydraulic mechanism for raising and lowering the undercarriage of an aircraft and in the large jacks and ramps used to raise motor cars so that wheels can be changed or the car can be worked on from below.

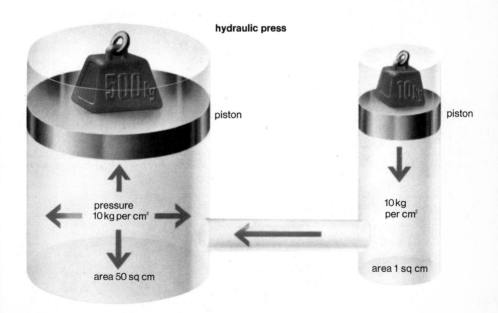

hydraulic press

piston

piston

pressure
10 kg per cm²

10 kg
per cm²

area 50 sq cm

area 1 sq cm

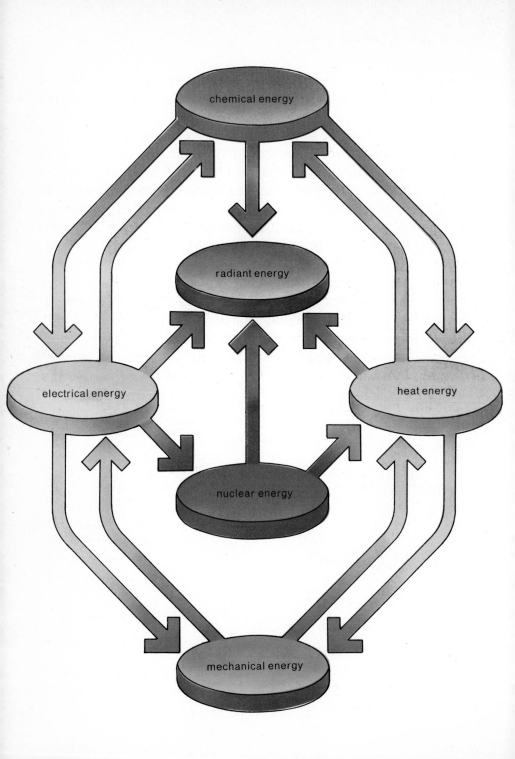

Energy is the ability to do work; to make things move. We usually think of energy as mechanical energy – the energy that moves a car or drives a machine in a factory. But where does the energy come from that makes a car run? It comes from the petrol in the tank: when it runs out of petrol, it runs out of energy and stops. The energy in the petrol is a form of stored energy, or chemical energy, because it is actually stored in the chemical bonds between the atoms that make up the petrol molecules. When the fuel burns inside the engine this chemical energy is converted into heat, and heat is itself another form of energy. The heat makes the gas in the cylinder of the engine expand and in doing so it pushes down the piston. This is an example of the conversion of heat energy into mechanical energy; it is the mechanical energy that is required to drive the wheels of the car.

An electric drill makes use of mechanical energy to turn the drill to bore a hole. The function of the motor in the drill is to obtain mechanical energy from the electrical energy supplied by the electricity power station. Thus the motor is a device for converting electrical energy into mechanical energy.

Several energy changes must take place in the power station to produce this electrical energy. If it is a conventional station using coal or oil the main source of energy is the chemical energy stored in the fossil fuel (coal and oil are fossilized remains of living plants and animals). This chemical energy is converted into heat energy in a furnace and raises steam. The steam drives a turbine, the mechanical energy of which is converted to electrical energy in a generator. In a nuclear power station, the prime energy source is nuclear (stored within the nucleus of an atom) rather than chemical. This nuclear energy is converted to heat energy in a reactor, which then follows the same steam-turbine-generator cycle.

During the 19th century the idea of energy began to crystallize and it was soon realized that there are several different kinds of energy, which can be converted into each other. It is possible to convert most kinds of energy into other kinds if the right type of device or process is used. Radiant energy, such as light or radio waves, however, can only be converted into other forms of energy if matter is present.

The idea that energy can be changed from one form to another is hardly more than a hundred years old; before that scientists had only vague ideas about it. In the 18th century heat was thought to be a weightless fluid, which was given the name caloric.

Part of the mystery was unravelled at the end of the century by Count Rumford who supervised the boring of cannon. He noticed that during the boring the metal became so hot that it could boil a can of water. His first idea was that boring a hole in the metal simply let the caloric run out. But when he found that a slow drill still made the metal hot even when it was too blunt to make a hole, he decided that heat could not be a fluid and that heat and mechanical energy must be connected. This idea led to the conclusion that heat itself is a form of energy.

The next step was taken by an Englishman, James Prescott Joule, who set out to measure exactly how much heat was produced by a certain amount of mechanical energy. Joule's apparatus had rather a comical look with weights and pulleys and paddles, like a strange egg whisk. It was arranged so that when a known weight was allowed to

Left: *apparatus to determine the mechanical equivalent of heat.* Right: *James Prescott Joule*

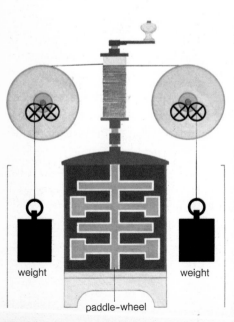

weight　　　　　　weight

paddle-wheel

fall through a measured distance, it turned the paddles which produced just enough heat in the water to raise the temperature of the water by a small but measurable amount. Like any other scientist, Joule repeated his experiment time after time: in every case the same amount of mechanical work produced the same amount of heat. In this way he not only proved that heat is a form of energy, but he actually measured the mechanical equivalent of heat.

As we now know that heat is just one of several forms of energy we do not have to consider the mechanical equivalent of heat. All forms of work and energy are measured in the same units, called joules. One joule is the work done when a force of one newton moves through a distance of one metre. (A newton is the force required to give a mass of one kg an acceleration of one metre per second per second.) In electrical units one joule is equal to the power of one watt acting for one second. So when you burn a 100-watt bulb for one hour you are using 100 watt-hours or 360 000 joules of energy. (One hundred watt-hours is, of course, equal to one-tenth of a kilowatt-hour, the unit in which electricity is sold.)

Left: *heat produced when boring a cannon.* Right: *Count Rumford*

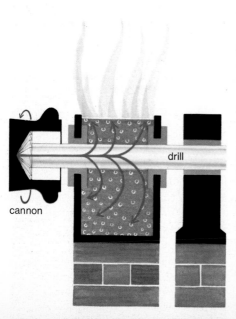

A motor car engine is a machine for converting the chemical energy stored in petrol into mechanical energy. Anyone who drives a car would like all the petrol to be converted into mechanical energy. If this could be done the engine would be 100% efficient. Unfortunately, this is only a dream because even the best-designed engine is unlikely to be more than 25% efficient. That is, only some 25% of the stored energy is converted into the useful energy required to drive the car.

It therefore seems as if 75% of the chemical energy has been lost. In fact, none of it is lost although a great deal of it is wasted. Some of the heat energy (a little less than 30%) is taken up by the water in the cooling system and wasted through the radiator to the air. Some energy escapes in the exhaust gases and some is wasted in the friction of the moving parts of the engine (between 40–50%). In addition, a certain amount of energy is never used because the petrol does not burn completely.

When energy is converted from one form to another, none of it is actually lost in the process; nor is any extra energy created. This fact is known as the law of conservation of energy. Wasted energy is

Fuel elements for a fast nuclear reactor

fuel value 100%

cooling water 35·8%

exhaust gases 35·6%

energy conversion in a car engine

different: whenever energy changes from one form to another, some of it is wasted in the form of heat. It is as if all energy is contained in a vast reservoir. But the reservoir has a leak in it so that heat is steadily leaking the energy away into an enormous pool of unusable heat energy. It will take billions of years for the reservoir to drain off into the pool: when eventually the reservoir is empty this will cause what is called the heat death of the universe.

But what about nuclear energy? We now know that energy and matter are two different forms of the same thing. In a nuclear reactor a tiny quantity of matter is converted into nuclear energy: this energy is not being created; it is being converted from matter. The law of conservation of energy is now known to be true only for systems in which there is no change of mass. It is matter and energy that are conserved, not just energy alone. The total quantity of all the matter and energy in the universe never changes, but they can be converted into each other according to Einstein's famous law $E = mc^2$, where E is energy, m is mass and c is the velocity of light. All nuclear reactors and atom bombs are based on this fundamental law.

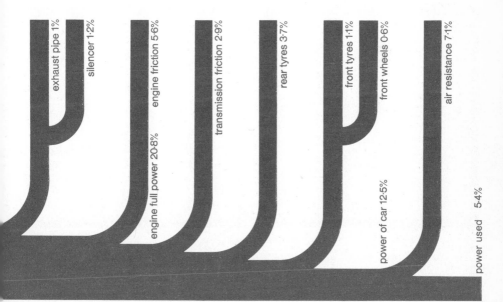

exhaust pipe 1% silencer 1·2% engine friction 5·6% engine full power 20·8% transmission friction 2·9% rear tyres 3·7% front tyres 1·1% front wheels 0·6% power of car 12·5% air resistance 7·1% power used 54%

All matter consists of atoms or molecules, which are small groups of atoms. The behaviour of atoms and molecules is simpler in gases than it is in liquids or solids. In the 19th century, scientists put forward the kinetic theory of gases to explain their behaviour. The theory assumes that all gases are made up of atoms or molecules, which are in constant and rapid motion. The atoms or molecules are continually bombarding the walls of the vessel which contains them and in doing so they exert a pressure on the walls. This pressure is entirely due to the momentum of the moving atoms or molecules.

The more heat a gas contains, the faster the molecules move. The faster they move, the more frequent and powerful are their collisions with the walls. This is why heating a gas increases its pressure. The heat of a gas is equal to the total of the energy of motion (kinetic energy) of all its molecules. Therefore the more gas molecules there are in a vessel the greater the quantity of heat it contains. Temperature, on the other hand, is a measure of the average kinetic energy of the molecules, so the temperature of a gas does not depend on the number of molecules present. The kinetic theory, therefore, makes a

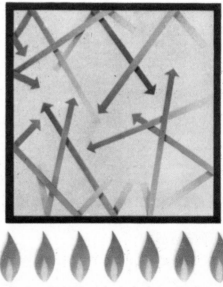

normal temperature and pressure   increase in pressure due to rise in temperature

very clear distinction between heat and temperature.

Once the ideas of the kinetic theory were accepted, all kinds of observations about the behaviour of gases could be explained. When a gas is compressed it becomes hotter. Notice how warm a bicycle pump becomes when you are pumping up the tyres. The pressure of a gas is due to the molecules bombarding and bouncing off the walls of the container. If one of the walls itself is moving inwards, in this case the piston of the bicycle pump, the molecules bounce off this wall with a greater speed than the speed with which they approached it. The energy of the moving wall is added to the energies of the molecules. Their kinetic energy is increased and so the temperature of the gas rises. In the diesel engine this rise in temperature when the piston compresses the gas is enough to explode the mixture of air and oil in the combustion chamber of the engine. In the petrol engine a sparking plug is necessary to make the mixture explode because the compression ratio is lower than that in a diesel engine. The higher compression ratio means that diesel engines must be considerably stronger and therefore heavier than petrol engines.

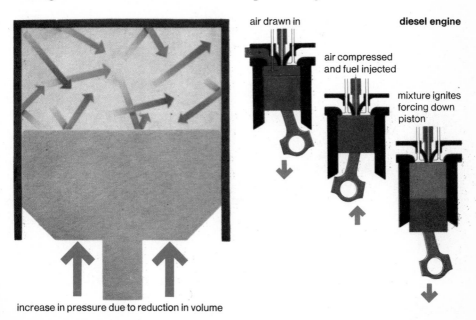

air drawn in

diesel engine

air compressed
and fuel injected

mixture ignites
forcing down
piston

increase in pressure due to reduction in volume

When a fuel burns, chemical energy is released in the form of heat. This is what happens when coal burns in a grate, although combustion of a fuel is only one example of a chemical change. Chemical changes always produce new substances and also, during chemical changes, there are always energy changes. Usually heat is given out and the change is then called exothermic. In some chemical reactions energy is absorbed while the new substance is being formed – this is called an endothermic change. To start a chemical reaction an external source of energy is nearly always needed. That is why chemists often heat their materials in test tubes or flasks to start them reacting. When you take a photograph the energy that starts the chemical change in the emulsion of the film is light, a form of radiant energy. This type of reaction is called a photochemical reaction.

Chemical changes can often be made to reverse. For example, oxygen and hydrogen combine to produce water; in reverse, water can be split into oxygen and hydrogen. When water is being formed energy is released, but the break-up of water absorbs energy.

When a chemical change takes place there is a rearrangement of the

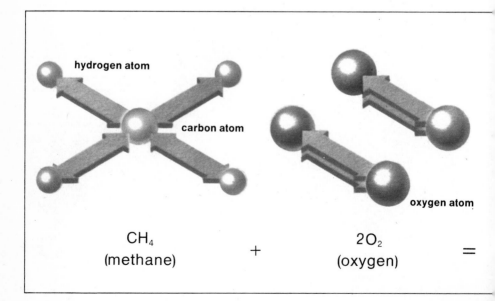

hydrogen atom

carbon atom

oxygen atom

CH_4
(methane)   + $2O_2$
(oxygen)   =

atoms and molecules. For example, methane the chief constituent of natural gas consists of molecules containing one carbon atom with four hydrogen atoms attached to it. The energy of the molecule is stored in the chemical bonds between these hydrogen and carbon atoms. When the gas burns, as it does in a gas oven, the methane reacts with the oxygen in the air and all the bonds between the various atoms break up. The oxygen atoms then form new bonds with the hydrogen atoms to make steam and with the carbon atoms to make carbon dioxide. The new bonds have less energy stored in them than the original bonds, so energy is given out during the reaction. The nitrogen in the air takes no part in the reaction; it only cools the hot gases. That is why pure oxygen gives a hotter flame.

Explosives are a rather special example of the way we make the release of chemical energy work for us. An explosive such as gunpowder or T.N.T. (trinitrotoluene) is a substance that splits up to form gases with the release of heat. The large volume of gases which are suddenly produced, and the heat which makes them expand, create the explosion.

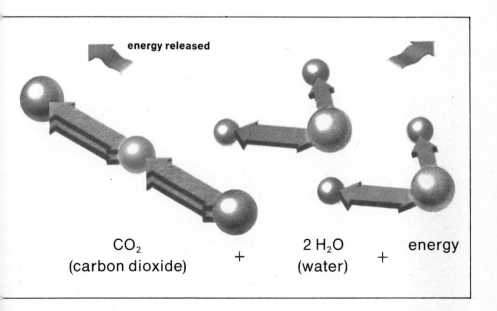

energy released

CO_2 + $2\,H_2O$ +   energy
(carbon dioxide)   (water)

All energy except nuclear energy comes from the sun. Not long ago it was thought that the sun was a giant fireball and burned like a coke furnace. But the sun is much too hot to burn. Matter burns when it combines with oxygen, releasing energy in the process. The sun is so hot that the form of matter we know on Earth cannot possibly exist. The temperature in the interior of the sun is so high that all molecules and most atoms would disintegrate into the tiny fragments called elementary particles. Only very stable atomic nuclei, such as helium nuclei (alpha-particles), can withstand the tremendous heat.

The interior of the sun is like an enormous H-bomb: hydrogen is being changed into helium and as it changes some matter is lost. This lost matter is converted into energy, which is then radiated out through space. Since the Earth is 150 million km away, only a minute fraction of this radiant energy ever reaches it. Yet more energy reaches the Earth every day than is stored in all the world's fossil fuels (coal, oil, natural gas).

The amount of solar radiation, called the solar constant, reaching the Earth can be measured. It is about 14 000 joules falling on every

Skylab orbiting the Earth

square metre of the Earth's surface in one second (about $\frac{1}{20}$ of the heat required to boil a kettle). From the solar constant it can be worked out how much energy is falling on a square kilometre of the Earth's surface. This is about 4000 megawatts (millions of watts) – sufficient to supply the power requirements of a small town.

One drawback to collecting this energy and converting it to a useful form is that the sun is not shining all the time; another is that the sunniest places on Earth are deserts where power is not needed. In spite of storage and transmission difficulties in turning solar energy into electricity, devices such as solar batteries for spacecraft have been developed. There are also ways of heating buildings from direct sunlight. Even in Britain there is sufficient sunlight to heat a house if well-designed solar heaters are built into the roof. But so far the most successful way to tap solar energy is to grow plants and then burn them. Coal and oil contain stored solar energy accumulated by plants over millions of years. Slightly more directly, water and wind power both come from the sun because it controls the weather on Earth.

The solar research establishment at Odeillo in the French Pyrenees uses a large parabolic reflector to collect energy from the sun

Since the Industrial Revolution, coal has been civilization's main source of energy. At the present rate of consumption, world supplies of coal will last for only about a hundred years. Its consumption, however, is decreasing rapidly as factories, ships and houses are converted to the use of oil. Oil reserves, too, are being used up and the energy crisis of the 1970s has shown that oil will become very expensive and hard to obtain by the end of the 20th century.

Coal is mainly fossilized carbon formed by the gradual decay of cellulose, the woody part of trees. Ordinary coal is about 85% carbon combined with hydrogen and small amounts of oxygen, sulphur and nitrogen. Burning is not only an inefficient way of using coal, but it also causes serious air pollution by producing soot and sulphur dioxide. If the coal is converted into coal gas in a gas-works not only is the gas burned but a smokeless fuel is produced in the form of coke. The impurities that would normally escape into the air can be converted into the many chemical by-products of coal.

Mineral oil is formed by the decomposition of marine plants and animals compressed under layers of sediment on the sea-bed,

how a refinery works

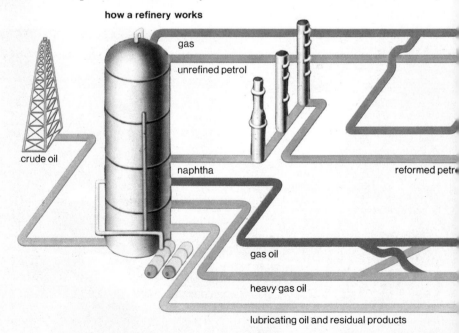

gas

unrefined petrol

crude oil

naphtha

reformed petro

gas oil

heavy gas oil

lubricating oil and residual products

although land movements may have shifted the deposits. Crude oil consists of compounds of hydrogen and carbon (hydrocarbons), which can be split up by distillation. The hydrocarbons that boil at a low temperature form petrol, while the higher-boiling ones give diesel fuel and heavy lubricating oils. Because of the enormous demand for petrol the larger higher-boiling molecules are also broken down into the smaller molecules of petrol by a process known as cracking.

Recently a new source of energy has become available in the form of the natural gas which collects over oil deposits. This gas can be released by drilling and is then piped to wherever it is needed. The main constituent of natural gas is the hydrocarbon called methane. Some of the gas used in British homes is natural gas from deposits under the North Sea.

Electrical power is being used to an increasing extent, but this too relies largely on coal and oil. About 85% of the electricity produced in the world uses coal or oil to heat the boilers which produce the steam for turbines which in turn drive generators. The remaining 15% of electricity comes from nuclear energy or water power.

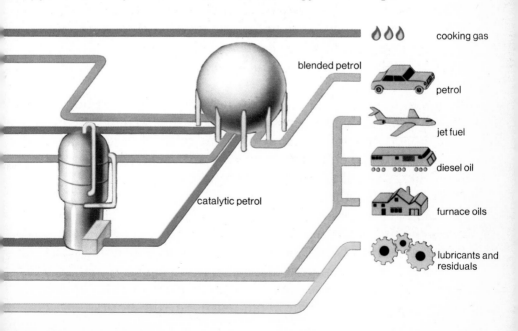

cooking gas

blended petrol

petrol

jet fuel

diesel oil

catalytic petrol

furnace oils

lubricants and residuals

Every plant and animal is a living machine: so is every individual living cell. As with any other machine, a living cell needs fuel to supply it with energy. The fuel for living organisms is food and like most other fuels food essentially consists of compounds of carbon and hydrogen, which react chemically with oxygen to release energy. These are typical combustion reactions, with carbon dioxide and water being formed exactly as they are when coal or oil burns in air. The main difference is that the combustion of food proceeds much more slowly than the combustion of a fossil fuel and the reaction is controlled so that energy is released as it is required.

This kind of reaction, in which a fuel burns in oxygen with a release of energy, will always continue once it has been started. It is much more difficult to make the reverse occur: only green plants can make carbon dioxide combine with water to produce such carbon compounds as sugar. Green plants are able to carry out this reverse process (photosynthesis) by using sunlight energy to sustain the reaction. All animals depend on plants for their basic energy supply: animals either eat plants or they eat the flesh of plant-eaters.

Sugar is the simplest food but there are many others, all of which contain carbon, hydrogen and oxygen. Bread and potatoes contain a large proportion of starch, which is a typical carbohydrate. The flesh of meat and fish is mostly protein – the second type of food. The third basic type is fat.

Just as engineers can work out the energy input and output of a machine, physiologists can calculate how much energy we need to take in, and how much energy different kinds of food will provide. The international unit for measuring heat and energy is the joule, but food energy is still measured in Calories, with a capital C: any diet sheet always has the energy content of food expressed in Calories. A Calorie is the amount of heat that will raise the temperature of one kilogram of water by one degree Celsius. An adolescent needs between 2500 and 3000 Calories a day; a grown man doing a desk job may need the same amount, but someone doing heavy manual work probably needs twice as much. A small piece of cake or one big potato would each provide about one hundred Calories.

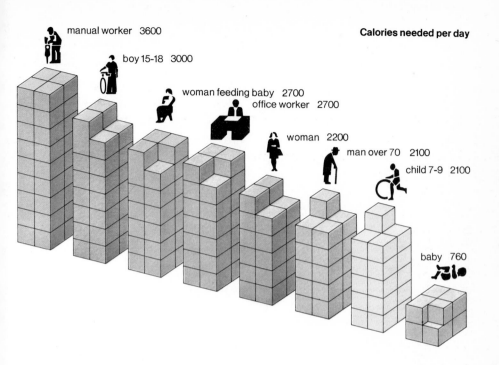

manual worker 3600

boy 15-18 3000

woman feeding baby 2700
office worker 2700

Calories needed per day

woman 2200

man over 70 2100

child 7-9 2100

baby 760

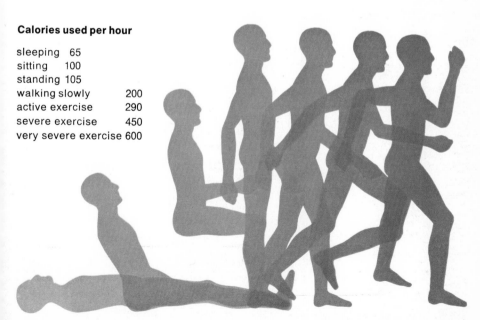

Calories used per hour

sleeping 65
sitting 100
standing 105
walking slowly 200
active exercise 290
severe exercise 450
very severe exercise 600

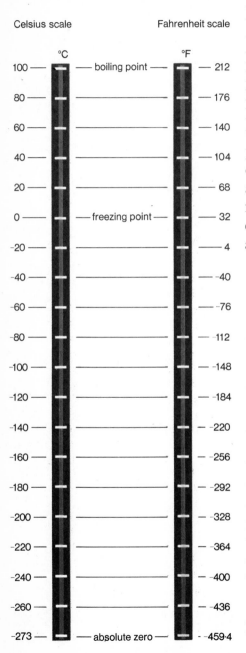

Celsius scale   Fahrenheit scale

°C   °F

100	boiling point	212
80		176
60		140
40		104
20		68
0	freezing point	32
-20		4
-40		-40
-60		-76
-80		-112
-100		-148
-120		-184
-140		-220
-160		-256
-180		-292
-200		-328
-220		-364
-240		-400
-260		-436
-273	absolute zero	-459·4

A red-hot tack is much hotter than a bowl of warm water. But the warm water contains more heat than the tack. This highlights the distinction between heat and temperature: the red-hot tack is at a higher temperature than the warm water. Temperature is a measure of the degree of hotness of a material and can be read on a thermometer. Heat is a form of energy. When a body is hot its atoms or molecules vibrate more vigorously than when it is cold. The heat of a body is a measure of the total of all the energy of motion of all its atoms and molecules. Its temperature is a measure of the average energy of motion of these atoms and molecules. There are more molecules in the bowl of water than there are in the small, red-hot tack (the water weighs much more than the tack). The water therefore contains more energy in its moving molecules than the tack. But the tack is at a higher temperature because the average energy of its molecules is greater than that of the water.

Temperature is measured in degrees. On a Celsius thermometer, which was devised by the Swede,

Anders Celsius, in 1742, the temperature of melting ice was originally taken as 100° and the temperature of boiling water as 0°. This scale was later inverted so that on the modern Celsius scale (Centigrade) water freezes at 0° and boils at 100°. The scale between these two 'fixed points' is divided into 100 degrees. On one of the earliest thermometer scales, devised in Germany by Gabriel Fahrenheit about 1724, the melting-ice temperature was marked as 32°F and the steam temperature as 212°F. This rather odd scale came about because Fahrenheit used a mixture of ice, salt and water to get the lowest possible temperature and then called it zero. His higher temperature was the temperature of the human body which he mistakenly thought was 100° on this scale (in fact it is 98·4°).

A low-temperature body may contain a considerable amount of useful heat – if its mass is large enough. In a large river such as the Thames the water is very slightly warmer than its surroundings because of the waste heat from the factories along its banks. As there are millions of tons of water in the river this low-temperature heat is worth removing and can be done by a heat pump.

Factories and hospitals near urban rivers can be heated by means of a heat pump

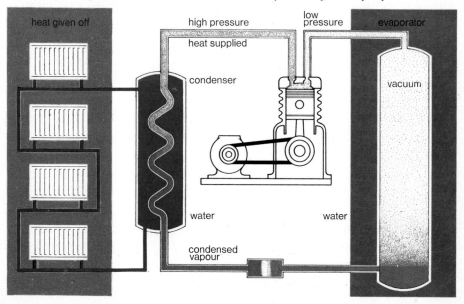

The usual way to measure temperature is by a liquid-in-glass thermometer. The liquid is either mercury or coloured alcohol. As the temperature rises the liquid in the bulb expands and moves up the tube which is marked in degrees. The range of temperature which can be measured on these instruments is limited by the freezing and boiling points of the liquids. Mercury freezes at $-39°C$ and boils at $357°C$. Alcohol freezes at $-112°C$ and boils at $78°C$.

For higher temperatures other types of thermometers must be used. When a metal wire is heated its electrical resistance increases with temperature and it becomes a poorer conductor of electricity. This is because the electrons in the metal move about more energetically as the temperature rises. Thus it is more difficult to make them flow in one direction – an electric current is a flow of electrons in one direction. This is the basis of the platinum resistance thermometer which consists of a coil of fine platinum wire connected to an instrument to measure the resistance of the wire. This thermometer can measure temperatures from $-200°C$ to $1200°C$. Because electrical resistance can be measured very accurately this thermometer can measure very small changes in temperature. When two wires of different metals are joined together and the junction of the wires is heated, a small electric current flows. This arrangement is called a thermocouple. The hotter the junction the greater the current. The thermoelectric thermometer uses a thermocouple of platinum and an alloy of platinum and rhodium. It is useful for temperatures from as low as $-200°C$ to about $1600°C$.

To measure the temperature of a furnace, which may be as hot as $2500°C$, a pyrometer is used. One type is the disappearing filament pyrometer. A special type of electric lamp is viewed against the light from the furnace through a red glass filter. The brightness of the filament can be altered by changing the current flowing through it and when it is exactly the same colour as the light from the furnace it becomes invisible. The temperature of the lamp filament, which can be calculated from the current passing through the lamp, is now the same as that of the furnace. A similar device can be used to measure the temperature of the stars by their colour.

liquid-in-glass thermometer

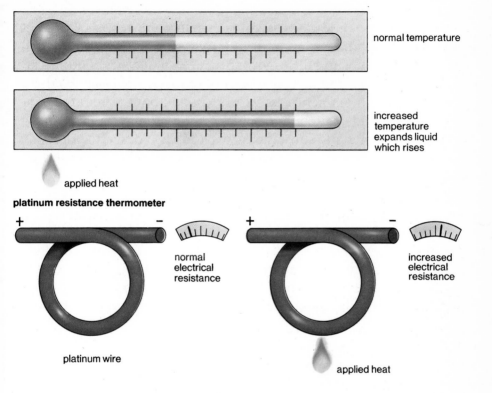

normal temperature

increased
temperature
expands liquid
which rises

applied heat

platinum resistance thermometer

+   −

normal
electrical
resistance

+   −

increased
electrical
resistance

platinum wire

applied heat

thermoelectric thermometer

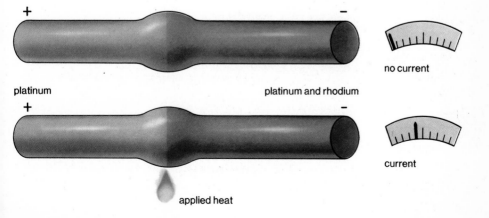

+   −

no current

platinum   platinum and rhodium

+   −

current

applied heat

Heat
63 expansion

One of the effects of heat is that it causes substances to increase (expand) in size. For example, the liquid in a thermometer expands. All solids, liquids and gases expand when they are heated. When heat is added to a solid, its atoms and molecules vibrate more energetically and therefore take up a little more space. The molecules of liquids and gases move faster and further when hot than when they are cold. For the same rise in temperature, gases expand about a hundred times more than liquids, and liquids about ten times more than solids.

Because different substances expand by different amounts, scientists can compare the expansion of substances by measuring a quantity called the coefficient of expansion. This is the amount by which unit quantity (length, area or volume) of the material will expand when heated through one degree Celsius. A bar of steel expands by 0·000 012 of its length when heated through one degree Celsius. This is an important fact for engineers. The Forth Bridge in Scotland is about one metre longer on a hot summer's day than in the cold of winter: if no allowance was made for this variation the bridge would be seriously buckled. The twisted steel girders of a building

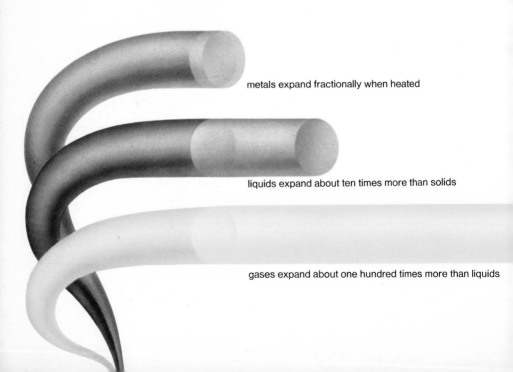

metals expand fractionally when heated

liquids expand about ten times more than solids

gases expand about one hundred times more than liquids

after a fire show the enormous forces that are exerted by expansion.

An alloy of steel and 36% nickel, called Invar, has a very low coefficient of expansion and is used in instruments when a variation in length with temperature would make the instrument inaccurate. If the pendulum of a clock expands the clock will go slower because the time of swing of a pendulum depends on its length – it is actually proportional to the square root of the length. For this reason pendulums and the balance wheels of watches are made of Invar. Expansion has useful applications in the measurement of temperature and also in devices called thermostats.

When substances are cooled they contract. But there is one exception to this rule: when water is cooled from 4°C to 0°C it expands slightly. It expands further as it turns into ice. This is why water pipes burst when water freezes. It is also why ice floats on water – as it expands it becomes less dense, and the less dense ice floats on the denser water. If ice did not float on water all the fish in a lake or pond would be killed when the temperature falls in winter; because it does float they are able to survive.

railway lines are set slightly
apart to allow for expansion

water boils over when heated

heated gases make balloons rise

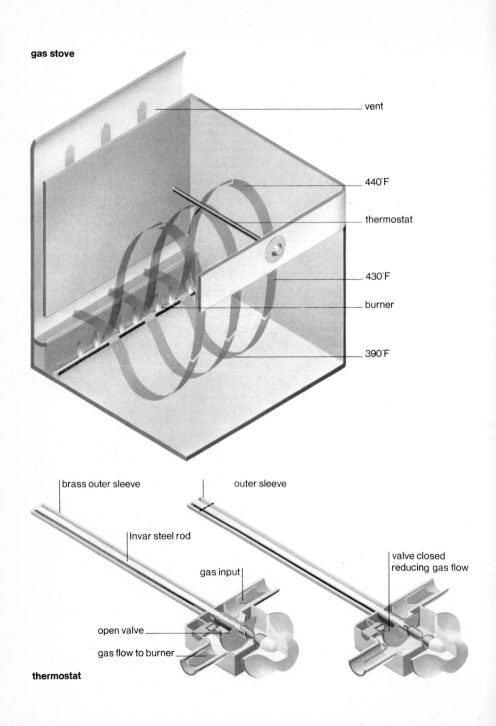

gas stove

vent

440°F

thermostat

430°F

burner

390°F

brass outer sleeve

outer sleeve

Invar steel rod

gas input

valve closed
reducing gas flow

open valve

gas flow to burner

thermostat

It is often necessary to maintain an even temperature. For example, on a gas or electric oven there is a dial that can be set to control the temperature and keep it constant. A device that maintains a steady temperature over a period of time is called a thermostat (from the Greek *thermo* meaning heat and *statos* meaning stationary). A thermostat must be both sensitive to temperature changes and linked to an electric control switch or a valve for supplying gas.

Refrigerators, water-heaters, electric irons, central heating systems and heated aquaria all have thermostats. Thermostats usually work by cutting off the heat supply when the temperature reaches a certain level and turning it on again when the temperature falls slightly.

Many thermostats depend on the fact that different materials expand by different amounts when they are heated. If two metal strips, one of aluminium and one of steel, are welded together a bimetallic strip is formed. When the strip is heated it bends because aluminium expands more than steel. The aluminium is on the longer, outside curve of the bend. A bimetallic strip can be made to work a switch and turn the electric current on and off. The gap between the switch contacts can be adjusted by a screw or dial to make the switching take place at different temperatures.

In a gas stove the thermostat consists of a brass tube with a rod of Invar steel running down its centre. The Invar steel has a very small expansion compared with brass. When the temperature rises the brass tube expands and cuts down the gas supply. The regulating dial alters the gap to be closed by the expanding brass and so the thermostat can be made to work at different temperatures.

Car engines also have thermostats, but they work on a slightly different principle. Instead of regulating the supply of fuel, they control the flow of cooling water between the engine and the radiator. When the engine is started from cold the thermostat cuts off the circulation of cooling water to allow the engine to warm up rapidly. As soon as the correct engine-running temperature is reached (usually between 70-80°C) the thermostat automatically opens and allows the cooling water to flow around the engine to keep its temperature constant.

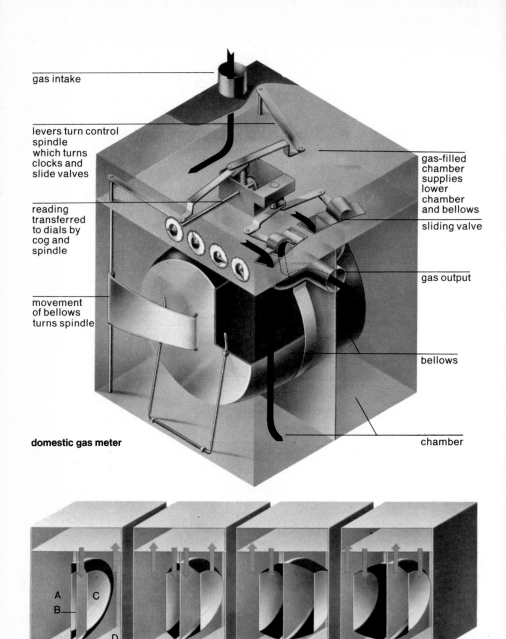

gas intake

levers turn control
spindle
which turns
clocks and
slide valves

gas-filled
chamber
supplies
lower
chamber
and bellows

reading
transferred
to dials by
cog and
spindle

sliding valve

gas output

movement
of bellows
turns spindle

bellows

domestic gas meter

chamber

A
B
C
D

Chamber A is full, bellows B empty; gas flows into bellows C forcing gas out of chamber D.
Gas flows into bellows B and C forcing gas out of chambers A and D. Bellows C full, chamber D
empty: gas flows into B and out of A. Gas flows out of chamber A and bellows C and into
bellows B and chamber D.

65 *measurement by energy units*

Since heat is a form of energy it can be measured in energy units. The unit used for measuring energy in the internationally-adopted SI units is the joule. One joule is equal to a power of one watt operating for one second. However, for some purposes other units of heat are used. The quantity of heat needed to warm something depends on the amount (mass) of the substance, what the substance is, and the rise in temperature produced. A well-known heat unit is based on this relationship and is defined as the amount of heat needed to warm one gram of water by one degree Celsius. This unit is called the calorie. A larger unit is the kilogram-calorie (often called the large Calorie and written with a capital C). This is the quantity of heat necessary to raise the temperature of one kilogram of water by one degree Celsius. When the energy values of food are calculated the large Calorie is always used.

Other heat units used are the British Thermal Unit (btu), which is the heat needed to warm one pound of water by one degree Fahrenheit (100 000 btus = 1 therm).

When we buy gas for our homes we pay for the amount of heat we are getting. So we pay so much a therm for it. Since there is no kind of meter to measure therms, the gas meter measures the volume of gas used in cubic feet and the Gas Board which supplies the gas has to guarantee that each cubic foot will produce at least a certain number of btus. This is called the declared calorific value of the gas. For natural gas it is about 1035 btu per cubic foot. So if you have used 5000 cubic feet of gas in three months, you will be charged for 5000 × 1035 = 5 175 000 btu or 51·75 therms.

Different substances require different amounts of heat to raise their temperatures by the same amount. For example, one kilogram of copper requires only about one-tenth the amount of heat required to warm one kilogram of water. The specific heat capacity of a substance is the amount of heat required to warm unit mass through unit rise in temperature. The specific heat capacity of mercury is 138·6 joules per kg per °C compared to 4200 J per kg per °C for water. One of the reasons that makes mercury a suitable liquid for thermometers is that such a small amount of heat is needed to warm it.

If one end of a poker is placed in the fire the other end soon becomes
too hot to hold. Heat has been transferred along the bar. This method
of heat transfer is known as heat conduction. The heat energy travels
along the bar warming up each section of the bar through which it
passes. Conduction is the passage of heat through a material from a
region of higher temperature to a region of lower temperature. Unless
there is a temperature difference no heat will flow.

All materials conduct heat to some extent but metals are the best
conductors. Scientists compare the ability of different materials to
conduct heat by their thermal conductivities (the amount of heat
flowing per second through a specified quantity of the material
against a specified temperature difference). Copper has a thermal
conductivity about a thousand times greater than that of glass and
about ten thousand times that of air. Poor heat conductors such as
glass, wood, plastics and air are known as heat insulators.

Conduction takes place when heat energy is transferred from one
atom or molecule of the substance to another. The atoms or molecules at the hot end of the bar have a greater kinetic energy and vibrate

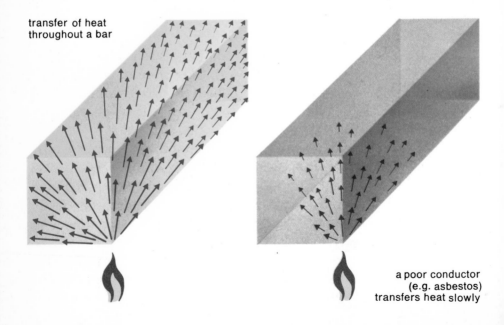

transfer of heat
throughout a bar

a poor conductor
(e.g. asbestos)
transfers heat slowly

more energetically than the cooler molecules. The vibration is passed on to neighbouring molecules, which in turn pass on the energy to their neighbours and then throughout the material.

To transfer heat from one place to another, for instance, from a gas flame to the water in a kettle, a good conductor of heat is needed. A copper kettle is better than one made of aluminium because copper has a higher thermal conductivity. Copper kettles are seldom used because of expense, but some kettles have copper bases.

To stop heat being transferred or lost a heat insulator is used. Hot-water pipes can be insulated by wrapping them with fibreglass or felt. These materials are themselves poor heat conductors but pockets of air are trapped between the loosely-packed fibres which add greatly to the heat insulating properties. It is the insulating effect of the air trapped in fabrics such as wool which makes clothes and blankets so warm. Fireproof suits are made of asbestos, a poor heat conductor, which is also non-inflammable. The hair of furry animals traps an insulating layer of air around their bodies which keeps them warm in winter and cool in summer.

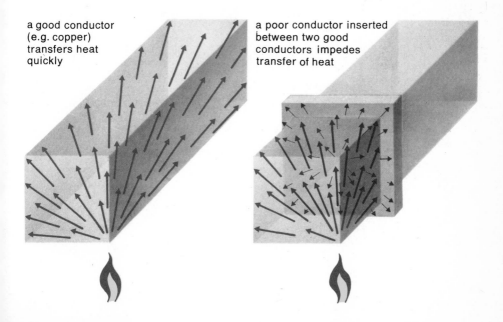

a good conductor (e.g. copper) transfers heat quickly

a poor conductor inserted between two good conductors impedes transfer of heat

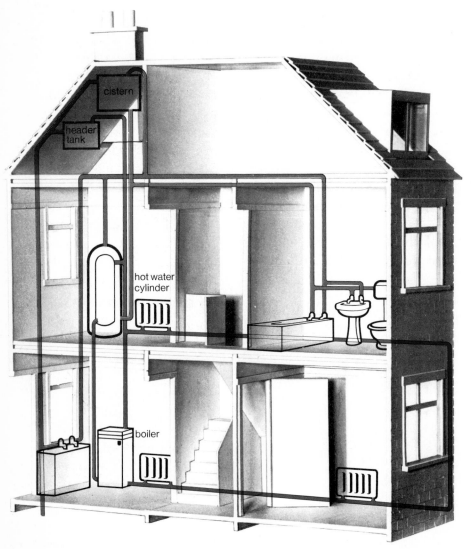

cistern

header
tank

hot water
cylinder

boiler

■ cold water
■ hot water
■ central heating

The water in a pan on a gas ring is heated all through, although heat is only supplied at the bottom. This is because the water circulates in the pan as a result of convection. The water next to the heat source becomes warm and expands. This water is then less dense than the colder water, so it rises, and the cooler water sinks to take its place. These water movements continue until all the water is at the same temperature. Convection currents occur in all fluids (liquids and gases) that are heated or cooled more in one place than another. Warm air rises and cold air sinks. On a small scale, convection currents are used in warming and ventilating buildings; on a grand scale they are the cause of winds and ocean currents.

A simple example of large-scale convection currents are the winds that occur in coastal areas. During the day the land absorbs the sun's heat more quickly than the sea. The air over the land therefore becomes warmer than the air over the sea; this warmer air rises and cooler air from over the sea moves in to take its place. This causes a breeze from sea to land – a sea breeze. At night the situation is reversed. The land cools down more quickly than the sea. The air over the sea is warmer and rises. The result is a land breeze, caused by the cooler land air moving out towards the sea. The wind systems of the world are mainly due to currents of hot air rising over the equator and cooler polar air moving in to take their place.

The domestic hot-water system is another example of the application of convection currents. In a simple system, water that has been heated in a boiler rises into a storage tank and cold water from a roof tank flows down to the boiler to be heated. The hot-water taps are fed by gravity from the storage tank. The central heating system works on a similar principle, except that when a number of radiators have to be supplied with hot water, the natural convection currents are often helped by a pump to maintain a good circulation. This is an example of forced convection. The water pump in a car helps the natural convection currents to circulate the water more quickly. The water is heated when it passes through channels in the cylinder block, thus keeping the engine cool. The water is then pumped through the radiator where it gives up its heat to the atmosphere.

If you stand around in wet clothes you will feel chilled because the evaporation of the water causes cooling. In order to evaporate and become a gas, the molecules of a liquid require extra energy to break the bonds between them so that they can be free of each other. In the case of the water in your damp clothes, the extra energy is taken from your body which is why you feel cold. This added energy required to change a liquid into a gas is called the latent heat of vaporization. In the same way, heat is also needed to melt a solid so that it becomes a liquid: this is the latent heat of fusion. The latent heat is the heat required to change its state without altering its temperature.

The cooling effect of an evaporating liquid is used in a domestic refrigerator. A refrigerant liquid (such as Freon, a compound of carbon, fluorine and chlorine) is pumped through cooling coils (the evaporator) in which it expands (evaporates) and absorbs heat from the surroundings. The evaporator is formed into the ice-making compartment of the refrigerator. After passing through the cooling coils in the evaporator, the vapour is then compressed by a compressor (usually driven by an electric motor) and condensed back to a liquid when the absorbed heat is given out. The cycle of events is then repeated over and over again. The refrigerator is really a heat engine working in reverse. In order to take heat out of the low-temperature interior of the refrigerator and transfer it to the higher temperature of the surrounding air, work must be done. If it is to work continuously, a refrigerator must be supplied with energy from outside. This external energy is usually electricity, which operates the electric motor driving the compressor, but it could be a gas flame. In the food chamber of a domestic refrigerator the temperature is just above the freezing point of water, about 1° or 2°C: in the ice-maker and in the deep-freeze it is usually around −15°C.

Modern methods of refrigeration have solved many of the problems of preserving fresh foods for long periods. In quick-freezing, the temperature of the food is dropped suddenly to below −30°C so that only very small ice crystals form. These do not damage the cell walls of the food, which retains its flavour and nutritious value. If food is cooled slowly, larger ice crystals form causing damage to the cells.

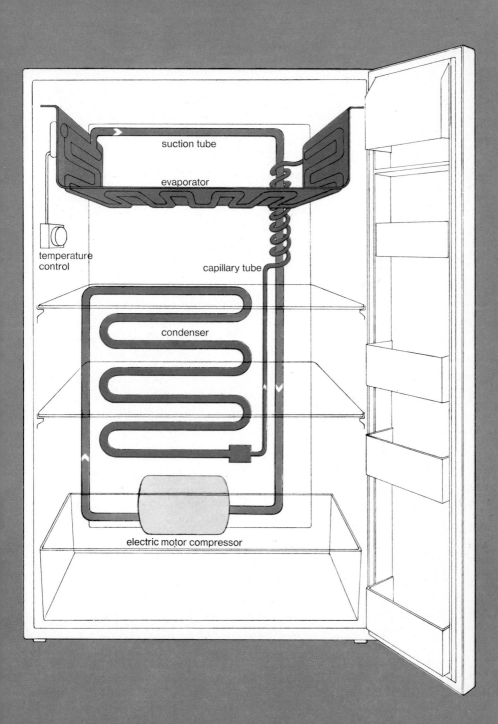

suction tube

evaporator

temperature control

capillary tube

condenser

electric motor compressor

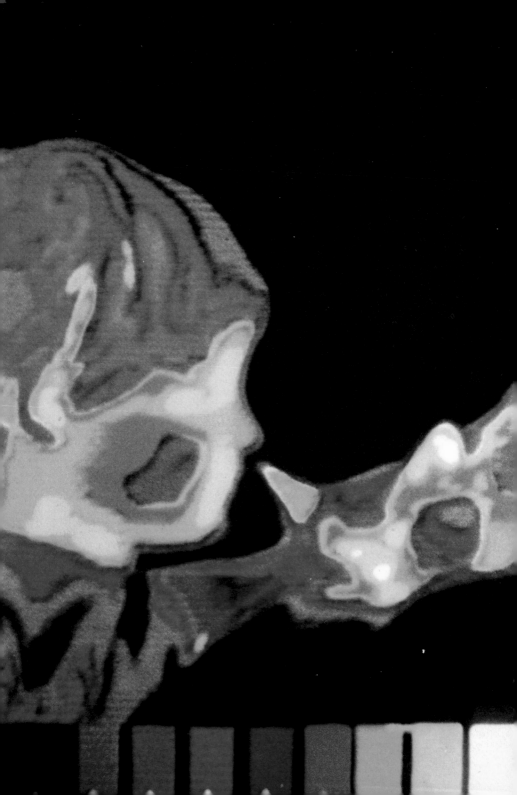

The Earth is warmed by the heat energy coming to it from the sun. This heat cannot travel from the sun to the Earth by conduction or convection because the space between the Earth and the sun is almost a vacuum. Conduction and convection can only take place where there is matter. The heat energy that travels through empty space is called radiant heat or infrared radiation.

In an eclipse of the sun the heat and light are cut off at the same moment: this shows that heat and light travel at the same speed. Radio waves also travel at this speed, and they too can travel through a vacuum. Radiant heat, light and radio waves are all forms of electromagnetic radiation: they differ only by their frequency (the number of complete cycles in one second). The frequency of radiant heat is less than red light, which is why it is called infrared, but it is greater than radio waves.

In the same way as light, radiant heat can be reflected, absorbed or transmitted by matter. Shiny metals reflect heat, but dull black surfaces absorb it. Waves of heat radiation only give up their energy when they are absorbed by an object. The object is then warmed, as we are warmed by the sun's rays. Most gases are transparent to radiant heat: glass is transparent to short-wave heat radiation but absorbs the longer waves. This is why a greenhouse acts as a heat trap. The glass allows the short waves from the sun to pass in, but the longer waves radiated by the warm objects in the greenhouse cannot pass out again. Energy, therefore, accumulates in the greenhouse.

Infrared waves are produced by the vibrating atoms and molecules of the source of the radiation. The higher the temperature of the source, the more rapidly the atoms vibrate and the greater the frequency of the waves. When the element of an electric stove is heated it first gives out only infrared radiation or black heat, as it is sometimes called. Then, as it gets hotter, its atoms vibrate more vigorously and it gives out the familiar red light that tells us that the current is flowing. If it were heated more it would give out white light (a mixture of all colours) and finally, if it had not melted, it would glow with blue light.

A thermograph of a boy and his cat, using infrared waves to record heat variations

The temperature of a substance is a measure of the average kinetic energy (energy of motion) of its vibrating atoms and molecules. If the vibrations were to cease altogether the average kinetic energy would be zero. This point is called the absolute zero of temperature. On the Celsius scale it is $-273 \cdot 15°$.

No substance of any kind could exist at a lower temperature than absolute zero. In fact, no substance could even exist at this temperature. Nevertheless, scientists working in low-temperature research stations have managed to reach a temperature as low as a few thousandths of a degree above absolute zero.

In the 19th century the idea of an absolute zero of temperature developed through the study of gases. Gases expand when they are heated and contract when they are cooled. It was found that all gases behave in the same way and have the same coefficient of expansion. Charles's Law, named after the French scientist J. A. C. Charles, states that the volume of all gases changes by $\frac{1}{273}$ of their volume at $0°C$ for each degree Celsius change in temperature, provided the pressure is kept constant. Thus if a volume of a gas at $0°C$ were

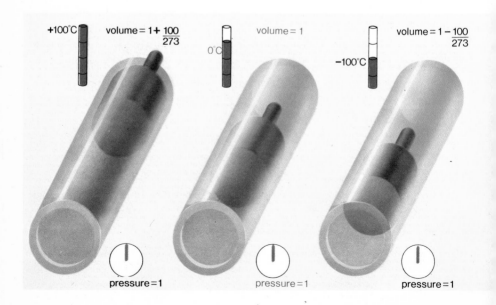

+100°C volume $= 1 + \dfrac{100}{273}$

0°C volume $= 1$

−100°C volume $= 1 - \dfrac{100}{273}$

pressure $= 1$ pressure $= 1$ pressure $= 1$

cooled down by 273° Celsius, it would contract by $\frac{273}{273}$ volumes, in other words its volume would be reduced to zero. Since it is impossible to imagine a substance that occupies no volume, a temperature of $-273°C$ is impossible to attain.

The study of the properties of materials at very low temperatures is called cryogenics. Some metals when cooled to within a few degrees of absolute zero lose their resistance to the flow of an electric current passing through them. In this condition the material is said to be superconducting. As there is no resistance, the passage of an electric current does not cause the material to heat up.

Because heat produced in the windings of an electric motor causes a serious loss of energy, motors are now being developed to run at these very low superconducting temperatures. Experiments are also being made to transmit electricity from power stations to homes and factories by underground cables, cooled to superconducting temperatures by using liquid helium. However, the cost of cooling the cables to these low temperatures may prevent an overall saving being made by this method of transmitting electricity.

Right: *liquid oxygen in a test tube*

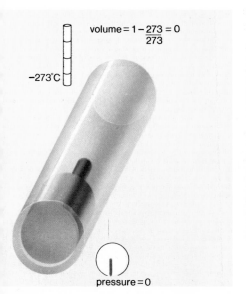

volume = $1 - \dfrac{273}{273} = 0$

$-273°C$

pressure = 0

Steam engines provided the first sources of mechanical power. Before they were invented man had to rely on his own muscle power or that of his domestic animals and the power of wind or water. The energy in a steam engine comes from the burning fuel which turns the water into steam. When water boils and becomes steam it expands over 1700 times and so can be made to exert a pressure to provide power.

Probably the first recorded movement produced by the expanding power of steam was a toy engine made by Hero of Alexandria about 200 B.C. who used jets of steam to make a metal ball revolve. Hero is also reputed to have made a hot-air engine which opened the doors of a shrine when a fire was kindled on the altar.

The first practical steam engine was invented in about 1712 by the Cornish blacksmith, Thomas Newcomen, who used it to pump water from tin mines. It was a very simple engine in which the pressure of the steam drove a piston up a cylinder. Then the steam was condensed by a jet of cold water while it was still in the cylinder, and the pressure of the atmosphere forced the piston down. Because the cylinder was alternately heated and cooled the engine was slow in working and

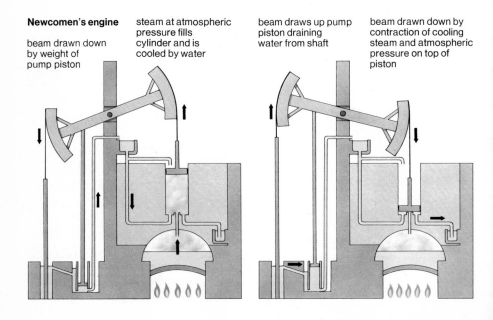

Newcomen's engine

beam drawn down by weight of pump piston

steam at atmospheric pressure fills cylinder and is cooled by water

beam draws up pump piston draining water from shaft

beam drawn down by contraction of cooling steam and atmospheric pressure on top of piston

most of the heat was wasted.

James Watt was instrument-maker to the University of Glasgow. In 1763 he was asked to repair a model of Newcomen's engine. This gave him the opportunity to study the principle of the steam engine and to apply his own inventive genius to the device. His first important contribution to the design was to add a separate condenser so that the cylinder did not need to be cooled at the end of every upstroke. This made the engine more efficient. Watt went into partnership with a Birmingham businessman, Matthew Boulton, to manufacture the engines. Later Watt invented the 'sun-and-planet' gearing which enabled the engine to turn a wheel and so drive machinery. About the same time another inventor used a crank for the same purpose.

Until 1787 the steam engine was a stationary engine. In that year the American inventor, John Fitch, produced the first steamboat, but it was not a financial success. By 1814 an English engineer, George Stephenson, had built the first practical steam locomotive to transport coal from the mines to the factories in Leeds. In 1829 he built his famous '*Rocket*' for the Liverpool-Manchester Railway.

An engraving from Mechanic's Magazine *in 1820 shows Stephenson's* 'Rocket'

The early steam engines were very inefficient. Watt's first engine had an efficiency of about 2%. This means that only one-fiftieth of the energy of the burning fuel could be converted into mechanical work. Even this engine was considerably more efficient than Newcomen's machine. Watt made many improvements on his own engines, but perhaps the most important advance he made was the invention of the double-acting engine. The steam was admitted by means of tubes and valves, first to one side and then to the other side of the piston. The steam pushed the piston backwards as well as forwards so that the engine no longer relied on the atmospheric pressure.

Since the aim of the engineer is to design as efficient an engine as possible, he needs to obtain all the mechanical work he can from the hot steam. Watt noticed that the exhaust steam was still at a fairly high pressure after it had moved the piston: it therefore still contained energy that was being wasted. In the first engines, steam was admitted to the cylinder during the whole length of the piston stroke. Watt now cut off the steam supply when the piston had travelled about a quarter of the way along the cylinder. The high-pressure

water   saturated steam   superheated steam   exhaust steam

steam continued to expand and push the piston along, saving valuable steam. In a modern engine the steam may expand as much as two hundred times. In a triple-expansion engine three cylinders are used, the exhaust steam from the first being led into the second. In this way as much heat energy as possible is extracted from the steam.

In the 1820s, the French scientist, Nicolas Carnot, worked out that the efficiency of a heat engine depends not on the properties of the working fluid but rather on the range of temperatures through which it works. In a steam engine the efficiency is equal to the difference in temperature between the hottest part (the steam) and its coldest part (the condenser) divided by the temperature of the hottest part. In modern steam engines, therefore, the steam temperature and pressure are made as high as possible. To obtain the highest temperature, water is boiled under pressures many times higher than the atmospheric pressure, and the steam is then passed through a series of hot tubes to superheat it before it enters the cylinder. Despite improvements made to steam engines, their efficiencies are still low (up to about 20%) compared with petrol and diesel engines (25–30%).

In a steam engine the backwards and forwards (reciprocating) movement of the piston must be converted by a crank into a rotating movement in order to drive the machinery. In windmills and water-wheels rotation is produced directly by the moving air or water acting on the blades of a wheel. But winds die down and streams dry up, making such devices unreliable. In the 19th century the use of steam in reciprocating engines suggested the possibility that steam could replace the wind – to make a steam windmill, or steam turbine. This can work in two ways.

A jet of steam can be directed onto paddles or buckets on the rim of a wheel similar in design to a water-wheel. A machine of this kind, called an impulse turbine, was invented by a Swede, C. G. P. de Laval, in 1889. Although it was extensively used for some time, modern turbines are reaction turbines. These work on another principle, which is familiar in the revolving lawn sprinkler. In a sprinkler it is the reaction of the jets moving forwards that drives the sprinkler backwards. The Englishman, Sir Charles Parsons, began to work on this idea in 1884, and all modern turbines are based on his invention.

Parsons' steam turbine

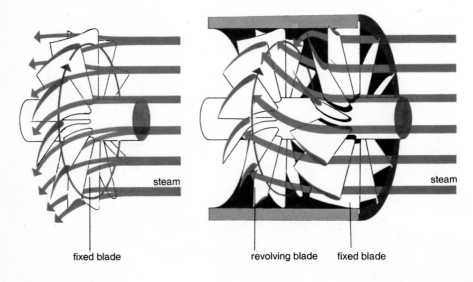

steam

steam

fixed blade   revolving blade   fixed blade

Parsons arranged a ring of fixed guide blades so they directed the steam onto a set of revolving blades. Then the steam passed through a second set of guide blades and onto another set of revolving blades attached to the same shaft as the first. In Parsons's first turbine there was a total of 25 pairs of blades. As steam expands its volume increases, so the diameters of the blade wheels were increased to allow for this expansion. In this way Parsons planned to extract a high proportion of energy from the steam. His first model was not very efficient, and engineers dubbed it the 'steam eater'. But modern improvements in design have made the steam turbine more efficient than the reciprocating steam engine. Turbines can be made larger and more powerful than reciprocating engines and are used in power stations and ocean liners. Modern steam turbines can achieve an efficiency of 25-30%.

The gas turbine uses the same principle as the steam turbine. Hot expanding gases from burning oil drive the rotor around. Turbines are used wherever high-speed continuous rotation is required. That is why the gas turbine is the basis of the modern jet engine.

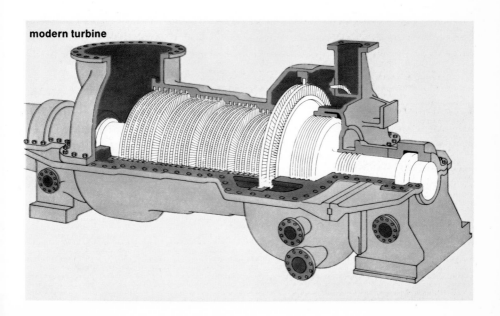

modern turbine

All heat engines operate by the expansion of gases. The energy to produce the expansion comes from burning a fuel. In steam-driven machinery the fuel is burned separately in the furnace of a boiler which produces the steam. The steam then enters the cylinder and moves the piston. In internal combustion engines the hot expanding gases are produced by burning the fuel inside the cylinder of the engine so that less energy is lost. These engines are generally smaller and more efficient than steam engines.

In the 18th century attempts were made to run engines by exploding gunpowder or burning turpentine in them. It was not until the end of the 19th century that easily vaporized oil products became available. These light oils (petrol) enabled a German engineer, Nikolaus Otto, to build the first practical internal combustion engine in 1876. The series of steps that takes place during the working of a petrol engine is known as the Otto cycle and is essentially the same today as when Otto developed it a hundred years ago. Two other Germans, Gottlieb Daimler and Karl Benz, perfected the design of Otto's engine and by mounting it on wheels produced horseless carriages, the first motor cars.

Petrol vapour and air are mixed, sucked into the cylinder where they are compressed, and then exploded by means of an electric spark. The resulting high-pressure gases force the piston down inside the cylinder. The burned gases are pushed out of the cylinder when the piston rises and the cycle is repeated over and over again. A movement of the piston up or down the cylinder is called a stroke and it takes four strokes (induction, compression, working and exhaust) to complete one complete Otto cycle during which the crankshaft turns around twice. Power is only produced during one of these four strokes (the working stroke).

Some smaller engines have a power stroke every other stroke; they do not work on the Otto cycle and are called two-stroke engines. They are much simpler as they do not have a complicated valve mechanism; the piston itself uncovers ports in the cylinder. In most modern car engines, however, four, six or eight cylinders, each working on the four-stroke principle, have their pistons connected to the same crankshaft and so produce a smooth power output.

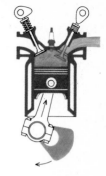

1 induction stroke   2 compression and firing   3 working stroke   4 exhaust stroke

Otto cycle

car engine

Many problems have to be solved in designing an efficient petrol engine. The liquid petrol must be vaporized and mixed with the correct proportion of air to form the best explosive mixture. The rate at which this mixture enters the cylinder has to be controlled so that the speed of the engine can be changed. Both jobs are carried out by the carburettor. A stream of air is sucked past a fine jet of petrol which causes the petrol to vaporize so that a petrol-air mixture is drawn into the cylinder. The butterfly throttle connected to the accelerator pedal acts like a tap to control the amount of mixture entering the cylinder. The proportion of air to petrol vapour is controlled by the size of the jet and the air inlet to the carburettor. To start a cold engine a mixture richer in petrol is needed, so the size of the air intake is reduced by using the choke. The carburettor is supplied with petrol from the petrol tank, either by a fuel pump, as in a car, or by gravity, as in a motorcycle.

The electric spark needed to explode the mixture must be supplied by the ignition system at exactly the right moment in the cycle. To produce the spark between the points of the sparking plug, the low voltage of the battery (12V) has to be converted into a high voltage (about 10 000V). This is done by the ignition coil which works like a transformer. In order to transform the voltage the direct-current supply has to be converted to an alternating supply by the contact breaker. Usually the contact breaker is inside the distributor, which supplies each sparking plug in turn with the high voltage at exactly the right moment in the cycle.

An internal combustion engine must be cooled. Only about a third of the energy of the exploding petrol is converted into mechanical energy – the rest produces waste heat. The heat must be removed to avoid damaging the cylinders. This is done in a car engine by a pump which circulates water around the cylinder block. The water is cooled in the radiator by a stream of air blown through it when the car moves. A fan increases the draught through the radiator when the car is stationary or moving slowly. The cylinder of a motorcycle engine has fins which distribute the waste heat into the air but as there is no fan, the engine will overheat if it is run too fast or too long without moving.

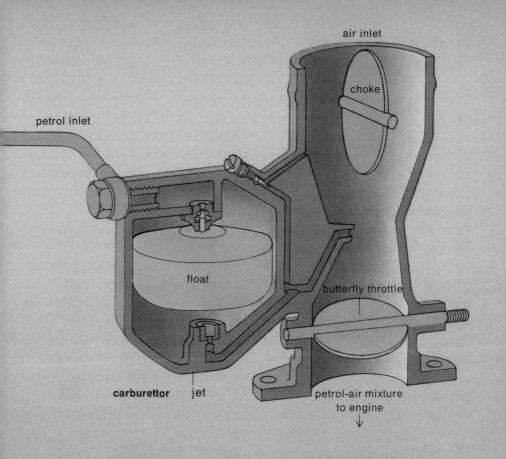

air inlet

choke

petrol inlet

float

carburettor jet

butterfly throttle

petrol-air mixture
to engine
↓

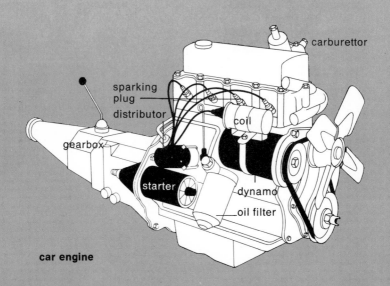

carburettor

sparking
plug

distributor

coil

gearbox

starter

dynamo

oil filter

car engine

The extent to which the petrol-air mixture is compressed in the cylinder of an internal combustion engine is called the compression ratio. It is equal to the ratio of the volume of the cylinder when the piston is at the bottom of its stroke to the volume of the cylinder at the top of its stroke. The more the mixture is compressed the hotter it becomes, and if it is compressed very rapidly the mixture will explode spontaneously without needing an electric spark to set it off. A German engineer, Rudolph Diesel, made use of this high compression ratio in his design for an engine which he patented in 1892.

Diesel compressed air in the cylinder to a very high degree, thus raising its temperature. Then the oil was injected as a fine spray into the cylinder just before the top of the upstroke: this raised the temperature of the oil to its ignition temperature (about 540°C) so that it exploded without the need of a spark. The four strokes of the diesel engine are (1) the air induction stroke when pure air is drawn into the cylinder, (2) the compression stroke when the air is compressed about twenty times, producing a temperature of over 550°C, (3) the working stroke when a fine spray of oil is injected into the

1 air induction

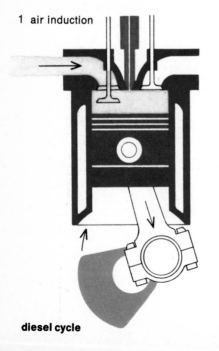

2 air compression
and fuel injection

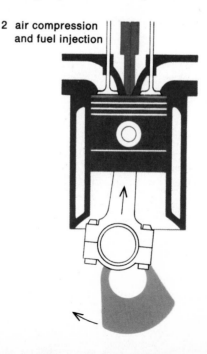

diesel cycle

cylinder so that it explodes and (4) the exhaust stroke to remove the burned gases.

Diesel's first engine, built in 1893, exploded and nearly killed him, but by 1897 he had made a successful engine producing 25 h.p. The diesel engine is heavier than the petrol engine because it has to withstand higher pressures, but it has no electrical system or carburettor and runs on heavier oil which the petrol engine cannot use. The efficiency of an engine is the energy output divided by the energy input (multiplied by 100 to make it a percentage). The efficiency of the diesel engine is about 35% compared with about 25% for the petrol engine.

Buses and many taxis use diesel engines. Large bulldozers have 400 h.p. diesel engines while marine diesel engines can develop up to 20 000 h.p. Diesel-electric engines, now used on some railways, are diesel engines which turn a generator which in turn supplies power to an electric motor. This arrangement is used because the diesel engine is most efficient when it runs at a steady speed. Electric motors are very flexible and do not require a gear box. The combination makes use of the best features of both types of motor.

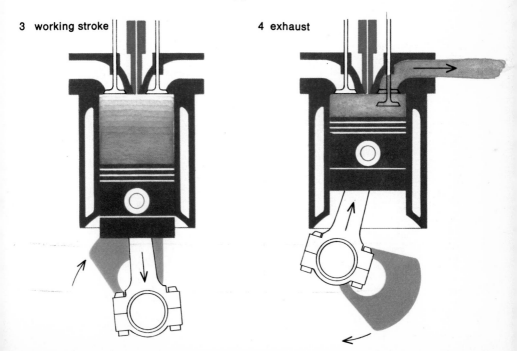

3 working stroke

4 exhaust

Engines are chiefly used to turn wheels or shafts, as in motor cars, trains and the machinery in factories. Most machinery works with a rotary movement, so the backward and forward stroke of the piston in a reciprocating engine must be converted into a rotary movement. The crankshaft and connecting rod system involves extra moving parts and thus more friction in the bearings that connect them. This reduces the efficiency.

One solution to this problem is the turbine, but gas turbines have a very high fuel consumption when used in cars. In an attempt to obtain the benefits of both piston engines and turbines a German inventor, Felix Wankel, produced in 1948 a new type of engine that bears his name. It uses the four strokes of the Otto cycle, but the piston rotates instead of moving backwards and forwards. There is only one moving part, the rotating piston, which is connected directly to the driving shaft. The valve mechanisms and crankshaft of the ordinary engine are unnecessary.

The Wankel engine consists of a shaped casing which houses a triangular piston with curved sides. This piston rotates so that the

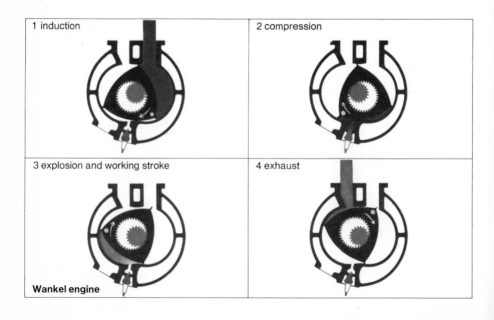

1 induction

2 compression

3 explosion and working stroke

4 exhaust

Wankel engine

corners, where there are special gas-tight seals, are always in contact with the casing walls. Thus three chambers are formed which vary in volume as the piston rotates. There is an inlet port admitting the petrol-air mixture from a conventional carburettor, an exhaust port and one sparking plug. In effect the engine has three cylinders, each of which goes through the induction, compression, working and exhaust strokes as the piston rotates. The working strokes for each chamber take place at different times.

The great advantages of the Wankel type of engine are in the small number of moving parts and lighter weight so that it has a greater power-to-weight ratio than the ordinary petrol engine. Because the engine has no reciprocating parts there is much less vibration than in a conventional engine. The Wankel engine also costs less to manufacture than a reciprocating type of the same power, and is now being used in several types of motor cars with some success. But as it still has its own problems, perhaps the most important of which is the provision of gas-tight seals at the corners of the piston, the Wankel engine is unlikely to replace the normal piston engine.

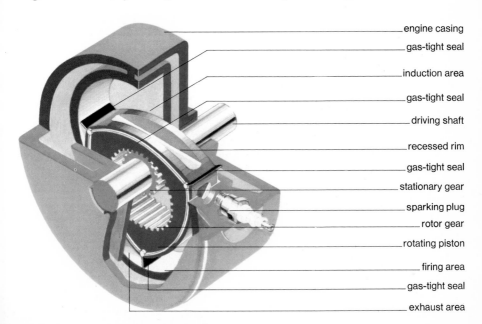

engine casing

gas-tight seal

induction area

gas-tight seal

driving shaft

recessed rim

gas-tight seal

stationary gear

sparking plug

rotor gear

rotating piston

firing area

gas-tight seal

exhaust area

Rockets and jet engines are heat engines in which hot expanding gases provide the driving force. Rockets were invented by the Chinese hundreds of years ago. Like today's modern firework rockets, they consisted of a paper cylinder filled with gunpowder which burned to provide a large volume of hot gas. The hot expanding gas supplied the driving force for the rocket in the same way as hot expanding steam drives a steam engine and hot expanding gases from burning petrol drive an internal combustion engine. Rockets work on the principle expressed by Newton's third law of motion – every action has an equal and opposite reaction. A rocket is simply a cylinder open at one end. When the fuel burns in the cylinder the hot gases expand in all directions and press equally on all parts of the inside of the cylinder. The pressures on the opposite sides of the cylinder cancel each other out so there is no sideways movement. There is a forward pressure on the closed end but no balancing pressure on the open end. So the rocket is pushed forwards. The forward force is called the thrust.

It is sometimes thought that the stream of moving gases from the

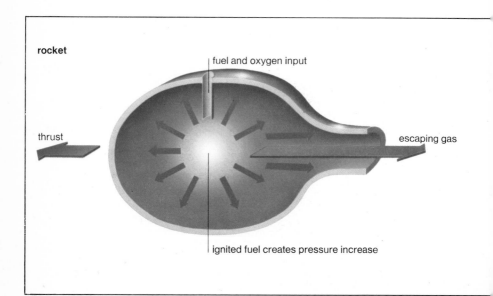

rocket

fuel and oxygen input

thrust

escaping gas

ignited fuel creates pressure increase

tail nozzle of a rocket forces the rocket forwards by pushing against the atmosphere. This is not correct, however: the gas stream would push the rocket forwards better in a vacuum, because there would be no air resistance to the motion. The atmosphere plays no useful part in rocket propulsion. This is why rockets are the only known method of propulsion in space as well as in the Earth's atmosphere.

A jet engine works on exactly the same principle as a rocket. The difference is that a rocket carries its own oxygen supply to burn the fuel while a jet engine uses oxygen from the air. This is why jet engines can only be used in the Earth's atmosphere. The simplest form of jet engine is the ram jet, consisting of a cylinder in which air enters at the front, fuel is injected and ignited and the hot gases expand outwards only at the rear end of the cylinder. The cylinder must be correctly shaped to ensure that the pressure in the combustion chamber is sufficiently high. Ram jets are so simple that they are often called 'flying drainpipes'. However, as ram jets must travel at a high speed in order to build up enough pressure to produce thrust, and they cannot take off by themselves, they are therefore of limited use.

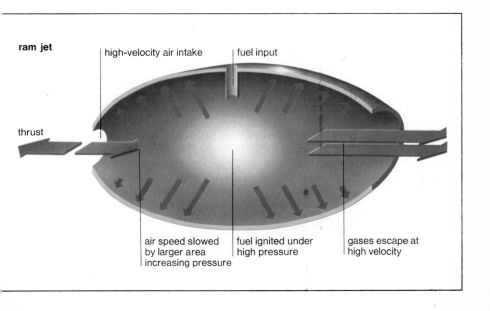

The simple ram jet relies on its forward speed to force air into the engine. This speed must be at least 500 km per hour if it is to be self-propelling. The flying bombs used by the Germans in the Second World War were crude jet engines powered by a type of ram-jet engine called a pulse jet. They relied on forward speed to build up the pressure in the combustion chamber, but combustion was not continuous; it was pulsed by air-intake valves at the front that closed when the pressure in the combustion chamber exceeded the pressure of the oncoming air. They therefore had to be launched by catapults.

Before jet engines could become practical for ordinary aircraft some way had to be found to make the aircraft able to take off without having to catapult it into flight. The solution was to fit a compressor to compress the air so that it entered the combustion chamber as fast as if the plane were flying at 500 km per hour.

The first jet engine for aircraft was the turbojet, designed in England by Sir Frank Whittle in 1941. In his model the burning gases first drive a turbine and are then expelled as a jet. The turbine drives the compressor which is on the same shaft. Because some of the

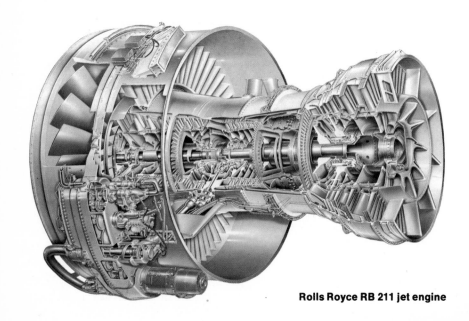

Rolls Royce RB 211 jet engine

energy of the jet stream is used to drive the turbine and compressor it is lost for providing forward thrust. At low speeds as much as 60% of the total power developed may be used to drive the compressor. The compressor itself is similar to the turbine and consists of a number of blades attached to a central shaft.

Whittle's turbojet engine has been modified and improved. Two or more compressors may be used, driven by two or more turbines for greater power. In turboprop engines the turbine drives a propeller as well as the compressor, and the jet stream provides very little forward thrust. In the by-pass turbojet some of the air from the compressor by-passes the combustion chambers but is channelled directly into the jet stream so that it increases the total mass of the exhaust gases and the engine is more efficient.

Jet engines are so successful that they have now taken the place of piston engines in all large commercial and military aircraft. They are efficient when flying at a great height, where the atmosphere is thinner, and are therefore most economical on long flights when there is sufficient time to climb to great heights.

turbojet aircraft

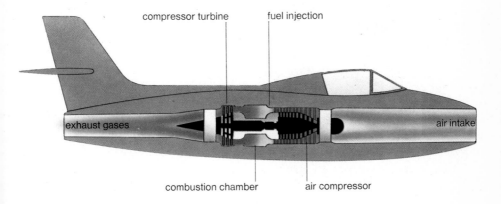

compressor turbine   fuel injection

exhaust gases   air intake

combustion chamber   air compressor

Space rockets must carry their own supply of oxygen to travel outside the Earth's atmosphere in addition to their own supply of fuel. The fuel may be solid, liquid or gaseous. A firework rocket is a solid-fuel rocket: the fuel is gunpowder, which supplies its own oxygen in the form of potassium nitrate (formula KNO_3, i.e. each molecule of potassium nitrate contains three atoms of oxygen). Many rockets use a liquid fuel, such as kerosene or alcohol, which is pumped, together with the oxidizer (hydrogen peroxide or liquid oxygen), into the combustion chamber at the rear of the rocket. There the two liquids burn, creating the large volume of hot gases that stream out of the rocket and produce the thrust.

An enormous force is needed to lift a rocket weighing several thousand tons off the ground and accelerate it to the speed of about 32 000 km per hour which is necessary to put a satellite into orbit. The exhaust nozzle of the rocket is shaped so that the jet stream moves directly backwards to produce a large forward thrust on the rocket. Gas which spreads out sideways does not contribute to the thrust.

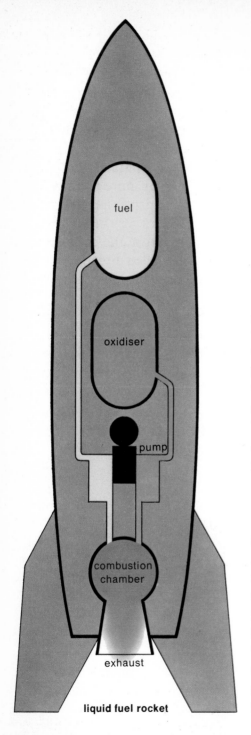

liquid fuel rocket

A rocket engine must withstand the high temperatures produced by the rocket fuels. Fuels which produce a temperature as high as 5000°C can be made, but steel melts at about 1300°C. However, new alloys with higher melting points have been developed and an efficient engine-cooling system designed for temperatures up to 3000°C. Most rocket failures are due to the engine itself being burned away.

The Saturn 5 rocket, which launched the Apollo spacecraft on its journey to the moon, weighed over 3000 tons at take-off. It was a three-stage rocket. The first stage, burning kerosene and liquid oxygen, lifted it 50 km above the Earth. This stage then dropped off allowing the remaining two lighter stages to be accelerated by the second stage to a height of 160 km. The third stage which, like the second stage, burned hydrogen and oxygen, thrust the spacecraft out of Earth orbit towards the moon.

Nuclear rockets are also being designed. They use a small nuclear reactor to heat a propellant gas, such as hydrogen.

Thor-Agena B rocket launches meteorological, communications and scientific satellites

Light
81   the electromagnetic spectrum

Light is a form of radiation to which the human eye is sensitive, and it is the electrical impulses sent to our brains which enable us to see. Without light we cannot see. The sun, stars, electric lights and flames all give out light – they are luminous. All the other objects we see reflect the light that falls on them. What exactly is light? In the 17th century the English physicist, Isaac Newton, suggested that it is a stream of particles, or corpuscles. This fitted in with everything that was known about light at that time, especially the way it travels in straight lines (rectilinear propagation) and is reflected by mirrors.

However, at the beginning of the 19th century another English physicist, Thomas Young, carried out an experiment that could not be explained in terms of particles. He showed that in certain circumstances two light beams can cancel each other out and produce darkness. This phenomenon, called interference, can only be explained if light is in the form of waves – darkness occurring when the trough of one wave coincides with the crest of another identical wave. Experiments on light beams with electric and magnetic fields now show that light consists of electromagnetic waves.

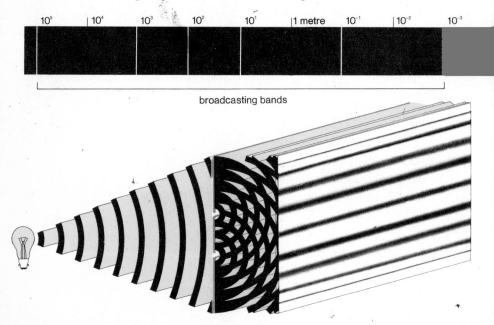

| 10^5 | 10^4 | 10^3 | 10^2 | 10^1 | 1 metre | 10^{-1} | 10^{-2} | 10^{-3} |

broadcasting bands

Electromagnetic waves are regular and rapid variations in electric and magnetic fields that spread out through space in all directions. The most characteristic property of a wave is its frequency – the number of crests or troughs there are in one second. Frequency is measured in units called hertz (one cycle per second).

Light is only one form of electromagnetic radiation and is only a small part of the complete electromagnetic spectrum. Radio waves and infrared waves are electromagnetic radiations with a lower frequency than light, while ultraviolet waves, X-rays and gamma-rays have a higher frequency. But all electromagnetic radiations travel at the same speed in a vacuum, about three hundred million metres per second.

The photoelectric effect – that light could release electrons from metals – could not be explained by the wave theory. Albert Einstein, in seeking an explanation, found it necessary to revert to a form of the corpuscular theory. We now combine the wave and corpuscular theories and think of light as a stream of small packets of electromagnetic waves called photons.

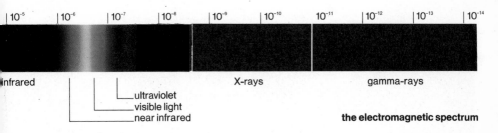

| 10^{-5} | 10^{-6} | 10^{-7} | 10^{-8} | 10^{-9} | 10^{-10} | 10^{-11} | 10^{-12} | 10^{-13} | 10^{-14} |

infrared   X-rays   gamma-rays

ultraviolet
visible light
near infrared

the electromagnetic spectrum

left: light passing through two slits in a board, projects an interference pattern onto a screen. The black stripes are produced when the light wave from the top source cancels out a light wave from the bottom source. White stripes result when waves from each source reinforce each other.

White light, or more accurately colourless light, from the sun or the white-hot filament of an electric light, consists of a mixture of light of several colours. The spectrum of colours which makes up white light ranges from the short violet and blue waves, through the green, red and orange waves to the red waves, which are the longest that the eye can detect. This was demonstrated by Newton in 1666. He allowed a narrow beam of sunlight to shine through a triangular glass prism and found that when the light that had passed through this prism fell onto a screen, it was split up into a rainbow-coloured band (a spectrum). Each component colour of the white light is bent, or refracted, through a different angle as it passes through the prism. In this way, the prism separates out the colours present in the white light.

To show that white light is a mixture of colours, it is necessary to combine all the coloured components to re-form white light. Newton's colour disc does this; it is a circular disc divided into sectors, each of which is painted with a colour of the spectrum. When the disc is rapidly rotated the colours are mixed in the eye and produce the impression of white. Mixing coloured beams of light in the correct pro-

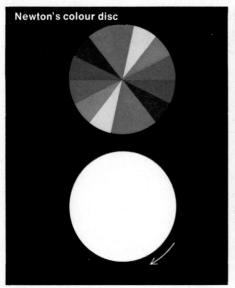

Newton's colour disc

the visible spectrum

portions of colours also produces a white effect. By changing the proportions, however, lights of many different colours can be produced.

Objects which are not themselves sources of light appear coloured because they reflect certain of the component colours of the light illuminating them to the eye. For example, a red book absorbs the violet, blue, green, yellow and orange components of the light falling on it and reflects only the red component. If all the light is scattered or reflected the object appears white; if an object absorbs all the light falling on it and reflects none, it appears black. Interesting effects can be produced by changing the colour of the light illuminating an object. A red book illuminated by a red light will appear red, but a red book illuminated with a pure blue light appears black because the blue light is completely absorbed and no light is reflected. Similarly, glass that appears red only allows red light to pass through it but all the other wavelengths are absorbed – it is transparent to red light only. The rainbow is a large-scale spectrum produced when the sun's rays pass through raindrops in the atmosphere. Each drop splits up the light in the same way as the prism in the spectrum experiment.

When light can easily pass through materials such as glass or perspex, they are called transparent. Other materials which absorb or reflect the light falling upon them are said to be opaque. A completely black surface absorbs all the light falling on it, while white paper reflects about 80% of the light which reaches it, scattering it in all directions. Highly-polished surfaces, a silver plate for example, reflect the rays falling on them in one direction only. These are mirrors and most ordinary mirrors are highly-polished glass backed with silvering.

A ball thrown against a wall will bounce off at the same angle at which it struck the wall. Light waves bounce off flat mirrors in exactly the same way. This is the law of reflection which says that the angle of reflection is equal to the angle of incidence. Mirror images are reversed so that when a page of print is held up to a mirror, the words go from right to left. When you look at your reflection in a mirror your right hand will seem to be your left hand and your hair parting, if on the left, will appear to be on the right. In all other ways, however, a flat mirror faithfully reflects the objects and scenes in front of it.

Mirrors change the direction of light waves and can be used to see

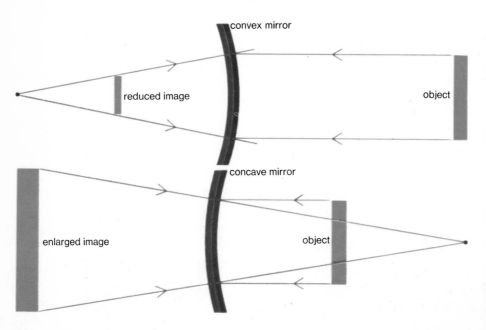

convex mirror

reduced image

object

concave mirror

enlarged image

object

around corners. A dentist uses a mirror to see behind the teeth. A car mirror shows what is behind on the road. With two mirrors set at suitable angles, which make a periscope, it is possible to see over obstacles. Two mirrors meeting at an angle form a set of symmetrical images resulting from multiple reflections, as in a child's kaleidoscope.

Light is reflected by curved mirrors in much the same way as by flat mirrors, but because each portion of a curved mirror faces a slightly different direction, the rays of light striking it are reflected at different angles. A saucer-shaped (concave) mirror reflects all light rays through the same point – this is called the focus. The image formed by a concave mirror can be the right way up or upside down, depending on the distance of the object from the mirror. It may also be either enlarged or diminished.

The opposite of a concave mirror is a convex mirror, which always gives an upright diminished image. Some car mirrors are convex to give a wider view of the road behind, making what the driver sees actually smaller than life-size.

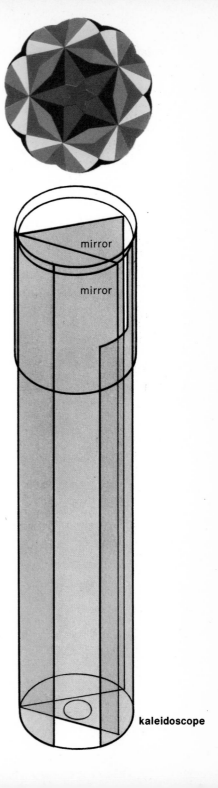

kaleidoscope

No matter what its source light always travels through a vacuum at a speed of about three hundred million metres per second. It also travels at very nearly the same speed in air, but when it encounters another medium, such as glass or water, it slows down. Different transparent substances slow light down by a different amount. The ratio of the speed of light in a vacuum to its speed in a particular medium is called the refractive index of that medium. The refractive index of glass is about 1·5, which means that light travels in glass at two-thirds of its speed in a vacuum.

When light passes from one medium to another, the change of speed alters its wavelength, making the light bend slightly (refraction). White light is split up into its separate colours on being refracted by a prism because each colour is refracted by a different amount. Refraction is also important in lenses. When a ray of light passes through a lens it is bent, and this enables the lens to produce the image of a distant object. The first lenses were made for spectacles about seven hundred years ago but it was not until about 1600 that scientists discovered that by looking through two lenses in line, small objects

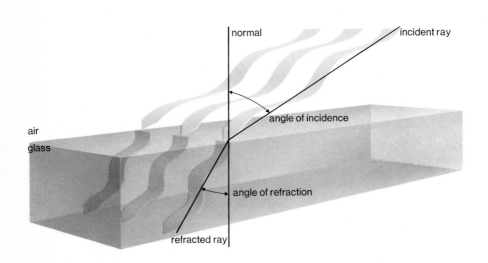

could be greatly magnified and distant objects made to seem close. This led to the development of the microscope and telescope.

A convex lens, which curves outwards and is thicker at the centre than at the edges, causes light beams to converge when they pass through it; a concave lens, which is thinner at the centre, spreads out the beam. Objects viewed through a concave lens always appear smaller. A simple magnifying glass is a double convex lens which causes a beam of light to converge to a focus (the point at which all the rays meet). When an object is fairly close to the lens, an upright magnified image is seen.

Objects seen through a cheap lens show a fringe of colours around the object which blurs the outline because the lens acts as a prism to split the light into a spectrum. This defect, called chromatic aberration, can be corrected by constructing a lens of two special types of glass. The outer part of a cheap lens also bends light more than the central portion, and gives the image a fuzzy look (spherical aberration). This effect is avoided in a camera by the use of a stop which lets light enter only through the centre of the lens.

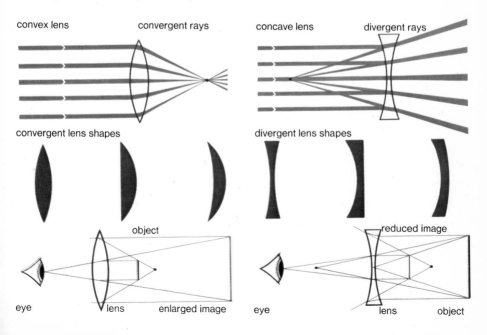

convex lens   convergent rays   concave lens   divergent rays

convergent lens shapes   divergent lens shapes

object   reduced image

eye   lens   enlarged image   eye   lens   object

A telescope is an instrument for magnifying distant objects and making them appear nearer. It has two essential parts: the objective, which collects light from the distant object and forms a real image, and the eyepiece, which forms a magnified image of this image. Refracting telescopes use a convex lens as the objective and reflecting telescopes use a curved mirror of large diameter.

A simple astronomical telescope for looking at the stars and planets consists of two convex lenses and produces an upside down image. For looking at distant objects on the surface of the Earth a terrestrial telescope is used which has a different kind of eyepiece arranged to produce a final image the right way up. One of the first telescopes, made by the Italian, Galileo, at the beginning of the 17th century, had a concave eyepiece and could magnify objects about thirty times.

Newton was the first to use a concave mirror for the objective of a telescope because he was dissatisfied with the colour defects (chromatic aberration) of images formed by large lenses. Mirrors are preferred to lenses because there is no chromatic aberration. Most modern astronomical telescopes are reflecting instruments based on

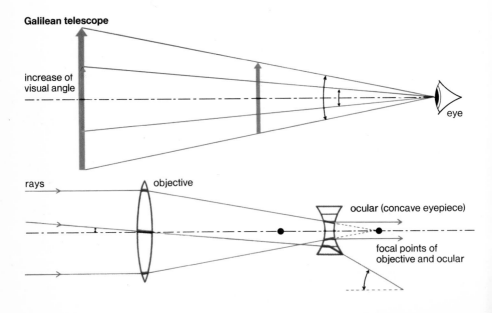

Galilean telescope

increase of visual angle

eye

rays

objective

ocular (concave eyepiece)

focal points of objective and ocular

Newton's use of mirrors. One famous reflecting telescope, completed in 1967, is named the Sir Isaac Newton telescope at the Royal Observatory at Herstmonceaux in Sussex. The mirror is 2·5 metres in diameter. The largest reflecting telescope in the world is at the United States Palomar Observatory in California. It has a mirror 5·08 metres in diameter and it makes stars appear a million times brighter than as seen by the naked eye.

The objective of a telescope should have as large a diameter as possible because the larger the diameter of the objective the more light will be collected, and so the final image will appear brighter. Also, the resolving power of a telescope increases as the diameter of the objective increases. The resolving power is the ability of an instrument to separate the images of two objects which are very close together, such as two stars forming a double star. It is because large mirrors can be made more accurately than large lenses that most modern astronomical telescopes are reflectors rather than refractors. The great accuracy with which mirrors can now be made reduces the amount of spherical and chromatic aberration.

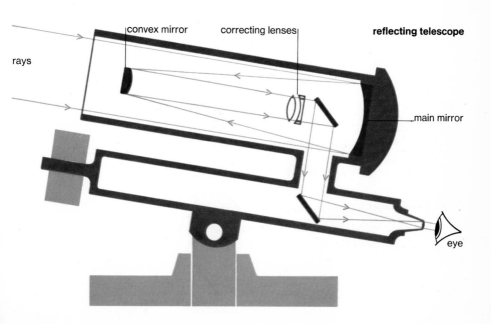

convex mirror   correcting lenses   **reflecting telescope**

rays

main mirror

eye

Microscopes are instruments that produce magnified images of very small objects. The first microscopes were simply magnifying glasses with devices that directed a strong light on an object. Around 1700 a Dutchman, Anton van Leeuwenhoek, made such excellent lenses that he was able to observe bacteria for the first time. This led to the important use of microscopes in modern medicine. At about the same time, in England, Robert Hooke developed a compound microscope with two lenses, very like those still used today. His instrument magnified a hundred times but a modern instrument with much better lenses can magnify up to about two thousand times.

The amount of detail that can be seen through a microscope is limited by the wavelength of light. This is because two points on a specimen cannot be distinguished from each other unless they are a certain distance apart. This distance must at least equal half the wavelength of the light used to illuminate the two points. The shorter the wavelength of the illuminating light the more detail can be seen. A microscope using ultraviolet illumination, quartz lenses and specially sensitive photographic plates can magnify up to three

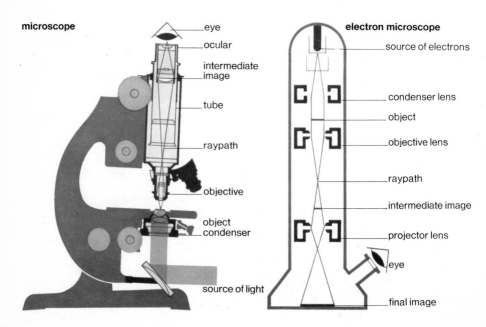

microscope

eye
ocular
intermediate image
tube
raypath
objective
object
condenser
source of light

electron microscope

source of electrons
condenser lens
object
objective lens
raypath
intermediate image
projector lens
eye
final image

thousand times. This is the greatest magnification that can be obtained by using glass lenses. The next step in the search for increased magnification was the invention of the electron microscope.

A stream of electrons behaves in many ways like light waves of very short wavelength. The faster the electrons move, the shorter the wavelength of the electron beam. A wavelength a thousand times shorter than visible light can be obtained by using a very high voltage to accelerate the electrons. It was this use of very short wavelengths, and so much greater magnification, that led to the development of the electron microscope. In an electron microscope the beam of electrons is produced by an electron gun, and the beam is focused by magnetic lenses in much the same way as glass lenses are used to focus a light beam. The final image is thrown onto a fluorescent screen or photographic plate because the eye does not respond to the extremely short electron waves. An electron microscope can produce magnifications of up to a million times and is used extensively to reveal the structure of minute objects such as viruses and the components of living cells that cannot be distinguished with a light microscope.

The tangled structure of a feather is revealed, magnified 650 times, by an electron microscope

 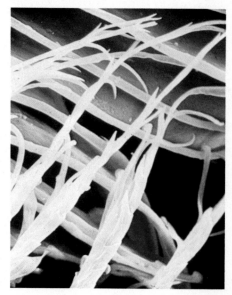

Many chemical changes need an external source of energy to start them. For example, a gas stove cannot be lit by simply turning on the gas: a spark is necessary to start the reaction between the gas and oxygen in the air. Heat is often used to start a reaction, but some reactions can be set in motion by light. A mixture of hydrogen and chlorine will explode to form hydrogen chloride when it is exposed to a bright light. This is an example of photochemical reactions. The most important photochemical change is photosynthesis.

Gases and vapours can be made to emit light either by heating or by bombarding them with a stream of particles such as electrons. The atoms of each element in the gaseous state have their own characteristic wavelengths at which they emit light. This pattern of wavelengths is called the emission line spectrum, because the spectrum is a series of lines instead of the continuous spectrum produced by white light from incandescent solids. When solids emit light, the atoms are so close together that the separate emissions interfere with each

Many advertising signs are lit by neon tubes

other and make a mixture of all wavelengths. This produces the continuous spectrum that is split up into a rainbow in a spectroscope. Emission line spectra detect the presence of a minute quantity of an element in a substance by examining the light the substance emits when it burns in a flame. This is called spectrographic analysis.

Neon signs are a familiar example of the emission of light by electron bombardment. A stream of electrons is passed through a tube containing the inert gas, neon, at low pressure. This stream of electrons causes the electrons in the neon atoms to jump to higher energy levels, at which time they are called excited atoms. When they return to their normal level they emit photons of the pink light which is characteristic of neon.

The photoelectric effect is the direct conversion of light into electricity. When light photons fall on certain metals they release electrons from the atoms of the metals. The number of electrons released depends on the intensity and colour of the light. Short wave photons have more energy than long wave photons and unless the photon has a certain amount of energy no electrons are emitted.

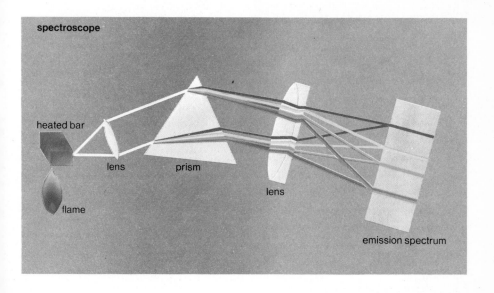

spectroscope

heated bar

lens   prism

lens

flame

emission spectrum

camera

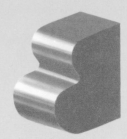

object to be photographed

developing

negative

enlarger

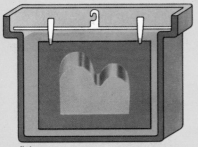

fixing

drying   negative

positive   print

Cameras were known long before men knew how to take photographs. In the 17th century Italian scientists devised a camera obscura – a dark chamber – with a convex lens that cast an image on a screen. It worked like a huge viewfinder and some painters used it to make exact drawings of buildings. In the 19th century French and English inventors recorded an image by means of a photochemical reaction on a glass plate – the effect of light on certain compounds of silver. Modern black-and-white photographic film consists of an emulsion of fine grains of silver bromide held by a thin layer of gelatin onto a strip of celluloid or cellulose acetate. After exposure to light, the film is developed in a chemical solution. This reduces the grains of bromide, onto which the light rays have fallen, to black metallic silver. The degree of blackening of the film, therefore, depends on the amount of light reaching a particular area.

The bright parts of the subject photographed emit or reflect much more light than the dull parts, and produce correspondingly darker images on the film. Light and dark are therefore reversed on the developed film which is why it is called a negative. To obtain a true reproduction, or positive, light is passed through the negative onto a light-sensitive paper, which also has an emulsion of silver salts on one side of it. The paper is then developed in the same way as the film until the desired contrast of light and shade is reached. The reduction process is then stopped by immersing the paper in a fixer, such as sodium thiosulphate (usually called 'hypo' because this chemical was formerly called sodium hyposulphite), and subsequently washing away the unexposed silver bromide. Black areas on the negative then appear as white areas on the positive and vice versa.

The amount of light entering a camera through its lens must be accurately controlled, by means of a shutter and a stop, to produce the required blackening on the film. The shutter, which is usually opened for only a fraction of a second, controls the exposure time. The amount of light entering the camera is also controlled by altering the diameter of the hole, or stop, behind the lens. In many cameras the distance between the lens and the film can be adjusted by a range-finder so that objects at varying distances can be sharply focused.

If the seven colours of the rainbow – red, orange, yellow, green, blue, indigo and violet – are mixed in the correct proportions, white light is produced. White light can also be made by mixing the correct proportions of only three colours, red, green and blue. These are called primary colours. By changing the proportions of any of the three primary colours, any other desired colour can be produced. Blue light mixed with red light gives a pinkish-purple colour called magenta. More surprisingly, red and green combined give yellow light. As white light is a mixture of red, green and blue lights it therefore follows that when yellow and blue lights are mixed they yield white light. Yellow is known as the complementary colour to blue. Similarly, magenta is complementary to green, and greenish-blue, called cyan, is complementary to red. Any coloured light is thus composed of different proportions of red, green and blue light. This is the basis of colour photography.

In black-and-white film there is a single light-sensitive layer, called the emulsion, that responds to light of any colour, but the final photograph is in black and white. There are three layers of emulsion in colour film. The top layer is sensitive only to blue light. Any blue light coming from the subject being photographed will produce an image on this emulsion. Blue sky or purple clothes will be recorded here, but not green grass or red brick walls. Similarly, green light will only affect areas on the second, green-sensitive emulsion, and red light will only be recorded on the lowest, red-sensitive emulsion.

After exposure to light, the film is developed and dyed so that a coloured image is produced on each layer of emulsion. Each of these negative images is in the complementary colour to that to which the particular emulsion is sensitive. A yellow image is therefore produced on the blue-sensitive emulsion, and so on. This image, however, is formed in regions not affected by the light. To form the positive image, white light is then passed through the film and the complementary colours combine to form the colours of the original subject. Yellow and cyan mixed give green, yellow and magenta give red, and magenta and cyan produce blue. These colours are reproduced in the appropriate areas on the final colour photograph.

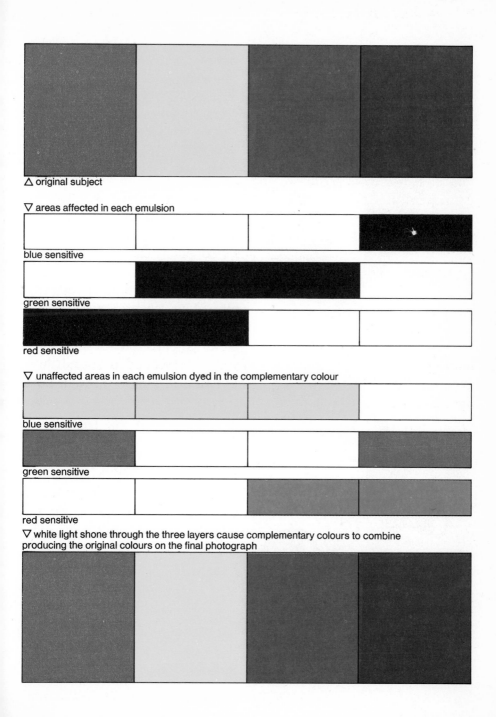

△ original subject

▽ areas affected in each emulsion

blue sensitive

green sensitive

red sensitive

▽ unaffected areas in each emulsion dyed in the complementary colour

blue sensitive

green sensitive

red sensitive

▽ white light shone through the three layers cause complementary colours to combine producing the original colours on the final photograph

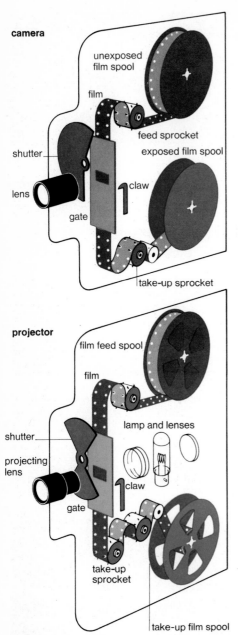

camera

unexposed film spool

film

feed sprocket

exposed film spool

shutter

claw

lens

gate

take-up sprocket

projector

film feed spool

film

lamp and lenses

shutter

projecting lens

claw

gate

take-up sprocket

take-up film spool

Motion pictures are a series of separate photographs taken at very short intervals. If the object in front of the camera is moving, each picture will be slightly different. When these pictures are subsequently projected onto a screen, there is an impression of continuous movement because each separate image is retained by the eye for a short time – this is called persistence of vision. To avoid a jerky image, the camera and projector should operate at a rate of at least 16 pictures per second. It is now usual to use 18 pictures per second for silent films and 24 per second for sound films.

The film moves through both camera and projector in a series of jumps. While the exposure is being made, or while the image is being projected, the film remains stationary after which it is quickly moved to the next picture. While the film is moving a shutter interrupts the light beam. The shutters in projectors usually have more than one blade making 36 or 48 light interruptions per second; this assists the illusion of continuous movement.

In normal cinematography, the pictures are projected at the same speed as they are taken. However, if the camera takes 120 pictures per second and the film is then projected at 24 pictures per second, the movements appear to be slowed down to one-fifth of their natural speed: this is slow-motion cinematography. When the taking rate exceeds about 1000 pictures per second it is high-speed cinematography. Slow-motion, which has many applications in science and industry, makes it possible to study movements too rapid for the eye to follow. It is also possible to use the reverse process to analyse very slow movements such as plant growth. For example, if one picture of a plant is taken every hour and the resulting film is projected at the normal rate of 24 pictures per second, the plant appears to be growing 84 000 times faster than it does in reality. Three days of growth would thus be shown in three seconds. This is time-lapse photography.

The sound track on film, which is either optically or magnetically recorded, is usually located in a narrow strip at the side of the picture images. The sound is thus synchronized with the pictures.

Time-lapse photographs, at 2·5 minutes per frame, show a mallow flower unfolding

At each end of the visible spectrum – the light to which our eyes are sensitive – there are bands of radiation that we cannot see. Infrared radiation, at one end, has longer wavelengths than red light, in the range of 0·4 millimetres to 700 nanometres (one nanometre is one millionth of a millimetre). Ultraviolet radiation, at the other end of the spectrum, has wavelengths shorter than violet light, in the range of 400 to 10 nanometres.

Infrared radiation, also called radiant heat, is produced by the vibrations of atoms and molecules. When a solid is heated its atoms and molecules increase their speeds of vibration and if the heating is continued long enough it begins to emit red light: it is red hot. But before a solid begins to glow it emits infrared radiation.

Infrared lamps are tungsten filaments working at a lower temperature (2500°C) than ordinary light bulbs (3000°C). They are used for quick-drying paints and inflammable liquids. The infrared radiation given off by hot objects can also be used to take photographs in the dark because photographic film can be made sensitive to infrared.

Forensic photography: concealed notes in this wallet are visible after treatment with anthracene, a fluorescing compound, and exposure to ultraviolet light

Infrared-sensitive films can detect cancer, as cancer cells emit more heat than normal cells: this is thermography.

Ultraviolet radiation, like light, is produced by the electrons surrounding an atomic nucleus when they jump from a high energy level to a lower energy level. About five per cent of the sun's radiation is ultraviolet and when these rays strike oxygen molecules (O_2) in the upper atmosphere some of these atoms are converted into ozone (O_3). The layer of ozone in the upper atmosphere then acts as a barrier which stops the penetration of most of the ultraviolet onto the Earth below. Without an ozone layer, life on Earth would be impossible because too much short-wave ultraviolet radiation damages living cells.

However, in small quantities the sun's ultraviolet radiation is valuable because it produces vitamin D. A narrow band of ultraviolet radiation, about 300 nanometres, also stimulates the production of melanin in the human skin. This is the brown substance that causes sun-tan. Ultraviolet lamps can produce a tan but they must be used for only very short periods or the rays will damage cells.

Infrared aerial view of countryside detects heat sources, including factories and houses

The word laser is formed from the initial letters of '*l*ight *a*mplification by *s*timulated *e*mission of *r*adiation'. Lasers produce a very narrow beam of light with all the waves of the same wavelength (monochromatic) and the same phase (coherent), that is, all crests of the waves occur at the samè instant.

Usually, when a solid is made to emit light by heating, the resulting light is a mixture of different wavelengths which are not in phase with each other. It is only by the process known as stimulated emission that monochromatic and coherent light can be produced.

Stimulated emission is a technique that has only recently been made possible, although it was predicted by the German-born physicist Albert Einstein in 1917. In a simple ruby laser a small cylindrical crystal of polished ruby, about one centimetre in diameter and five centimetres long, is silvered at one end and partly silvered at the other. The crystal is surrounded by a powerful spiral flash lamp. When the flash lamp is switched on, the chromium atoms in the ruby crystal are raised to excited states – the electrons jump a little further away from the nucleus. When they jump back they emit light. The

laser

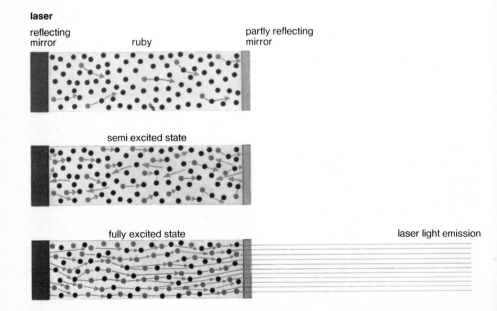

reflecting mirror   ruby   partly reflecting mirror

semi excited state

fully excited state   laser light emission

object is to make as many electrons as possible jump back to the normal state at the same instant to create stimulated emission. When a few electrons have spontaneously jumped back to normal levels, the light they produce is reflected up and down the crystal by the silvered ends. Each time one of these reflected rays strikes an excited chromium atom, the atom emits its own burst of light. An atom that has been stimulated in this way emits its light at the same wavelength, in the same direction and in the same phase as the stimulating light. It is rather like a burst of cheering at a football match when a few people cheer spontaneously and this stimulates the whole crowd.

The monochromatic coherent light is emitted in pulses from the half-silvered end of the crystal in a very narrow beam that is powerful enough to melt metals. Lasers can therefore be used in delicate industrial welding, in eye surgery to fuse detached retinas in place and in dentistry to drill holes in decayed teeth. Many other uses are being developed for these powerful devices. The latest forms can produce stimulated emission in low-pressure gases, such as neon and helium, as well as in semiconductors.

Lasers, a relatively new technical tool, have many uses in industry and medicine

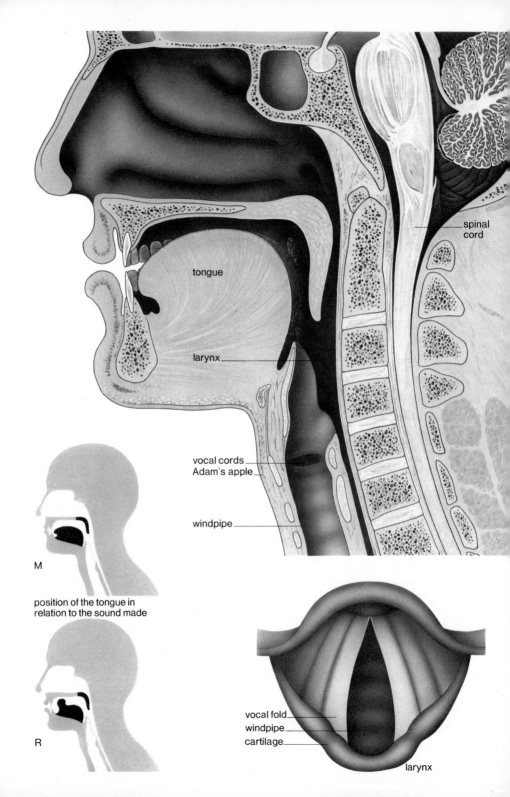

spinal cord

tongue

larynx

vocal cords

Adam's apple

windpipe

M

position of the tongue in
relation to the sound made

R

vocal fold

windpipe

cartilage

larynx

Sound

93 creating waves

Sound is always caused by something moving – the slamming of a door, the running of a car engine, footsteps, the wind rustling the leaves of a tree. Every movement sets up vibrations which cause changes of pressure in the surrounding air. Sound waves are created when these changes of pressure spread out in all directions, like the waves on the surface of a pond when a stone is tossed into it.

Sound waves cannot travel through empty space because they need a solid, liquid or gas to pass through. Astronauts on the moon are equipped with radios to speak to each other. Even if they were able to live outside their space suits, it would be impossible for them to talk to each other as they do on Earth, because the moon has no atmosphere through which sound waves could travel.

Sound waves can be produced in several different ways – by hitting, rubbing, plucking and blowing. A drummer hits the tightly-stretched skin of the drum. A violinist draws his bow over the stretched strings to make them vibrate. The strings of the harp and guitar are plucked. All wind instruments are played by blowing air through a pipe so that the air inside the instrument is forced to vibrate in a number of different ways.

The human voice is a wind instrument. Two thin pieces of skin, the vocal cords, are at the back of the throat. Breathing out between these stretched membranes makes them vibrate and produce sound waves. Different sounds can be made by using muscles to alter the tightness of the vocal cords. Then, by changing the shape of the mouth and the position of the tongue, the great variety of different sounds that make up everyday speech can be produced.

As everything that vibrates at a frequency of between about 50–20 000 vibrations per second makes sound waves, not all sounds are intentional. Accidental sounds that are loud or discordant are called noise. Car engines, aeroplane engines, road drills and factory machines all have vibrating parts and they all produce unwanted sound waves that serve no useful purpose. Apart from wasting valuable energy this noise is one of the urgent problems of urban life and a great deal of time and money is spent on the control and suppression of unnecessary noise.

A shrill high note is produced by rapid vibrations: a deep low one by slow vibrations. The highness or lowness of a note or sound is called its pitch and the pitch of a note depends only on the number of times the sound producer vibrates in one second. The number of vibrations in one second is called the frequency of the sound. When the key for middle C is struck on a piano, a hammer strikes two or three strings, each of which vibrates 256 times in one second. The modern unit of frequency is the hertz and one hertz is equal to one vibration or cycle per second: the frequency of middle C is therefore 256 hertz.

The human ear cannot detect very low- or very high-frequency sounds. It is sensitive to frequencies of from about 20–20 000 hertz but the range varies from person to person and it becomes narrower with age. Good musical reproduction requires a faithful reproduction of all sounds up to a frequency of 12 000 hertz, although reproduction of speech requires a much lower upper frequency.

To obtain different musical notes the frequency of the vibrations of the sound producer must be changed. This can usually be done by altering the size, the tightness or the weight of the object which is

decibel ratings of various sounds

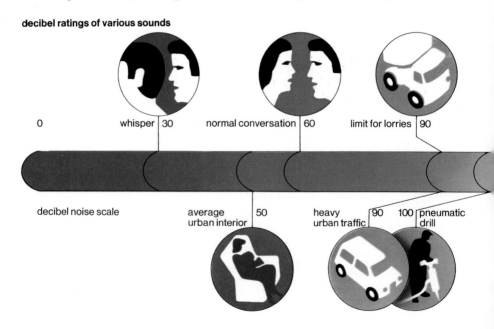

| 0 | whisper | 30 | normal conversation | 60 | limit for lorries | 90 |

decibel noise scale

average urban interior 50   heavy urban traffic 90   100 pneumatic drill

vibrating. If the length of an elastic band is altered while it is being plucked there is a change in pitch. A violinist adjusts the tightness, or tension, of the strings of his instrument in order to tune them. Then he alters the length of each string by 'stopping' them with his fingers to play different notes. In wind instruments the length of the vibrating air column must be altered to change the pitch of the note produced.

The loudness of a sound is the effect it has on the human ear. It depends on the size (amplitude) of the vibrations. The larger the vibration the louder the sound. The amount of energy needed to produce the vibrations controls the intensity of the sound. Although the pitch of a sound can be judged very accurately, our ears are not very good at judging loudness.

The loudness of sounds is measured in units called bels, or more often in tenths of this unit called the decibel. A whisper has an intensity of about thirty decibels, normal conversation about sixty decibels, and a jet aircraft 30 metres away has an intensity of 140 decibels. This is the danger limit for the unprotected ear, and people exposed to this level of noise should wear some form of ear cover.

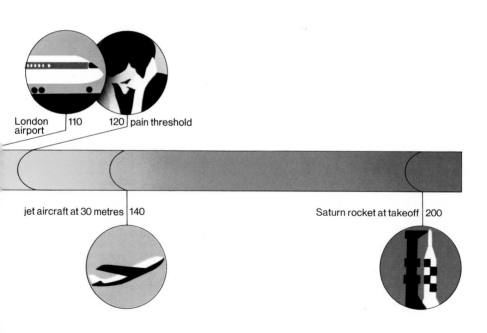

London airport 110   120 pain threshold

jet aircraft at 30 metres 140   Saturn rocket at takeoff 200

The pitch (which depends on frequency) and loudness (which depends on amplitude) of a note can be easily recognized. The third characteristic of a note is its quality or timbre, which enables us to distinguish between the same note of identical pitch and loudness played on different instruments. For example, middle C played on the piano sounds different from the same note played on the violin.

The quality of a note depends on the fact that very often, when a sound is produced, other notes of a higher pitch are also produced. When a tuning fork is struck, only one true note is produced – the fundamental – with a precise pitch and frequency. It can be illustrated as a simple smooth wave. The simplest way in which a string can vibrate is in a single loop which gives the fundamental note. But depending on where the string is struck or plucked other kinds of vibrations, or overtones, can occur. Overtones or harmonics have frequencies which are simple multiples of the frequency of the fundamental.

Overtones are created by the string vibrating in two, three, four or more loops. The double loop makes a note which is an octave higher

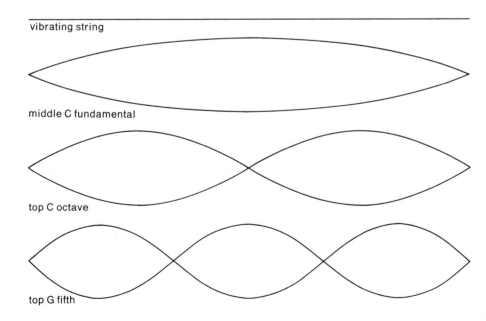

vibrating string

middle C fundamental

top C octave

top G fifth

than the fundamental and has double the frequency. The treble loop produces a note with three times the frequency of the fundamental, and so on. When a piano string is struck we hear the fundamental note plus several higher overtones. When a violin string is bowed we hear the fundamental note plus a different set of overtones which give a different quality. The quality of a note depends on the mixture of overtones it contains, each quality having a different wave form. Poor reproduction of sound in a radio set or record player is usually due to loss of overtones. High-fidelity record players have carefully balanced loudspeakers and electronic circuits that accurately reproduce the quality of the notes played.

A collection of notes whose frequencies are in simple ratios is a musical scale. The simple scale – doh, ray, me, fah, soh, la, ti, doh (diatonic scale) – has its notes in the ratios $1:9/8:5/4:4/3:3/2:5/3:15/8:2$.

The equitempered scale has 12 semitones in the octave. Each note bears the same frequency ratio ($\sqrt[12]{2} = 1{\cdot}0595$) to the next one to it. It is now used for most music written in the western world.

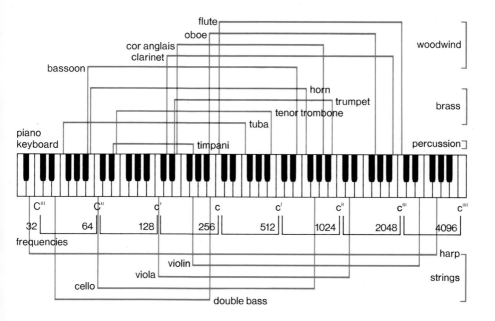

Sound waves seen on an oscilloscope screen show the different wave patterns of three musical instruments. The trumpet's sound wave makes a pattern that repeats itself with almost perfect regularity. This indicates a clear note which has constant frequency and intensity.

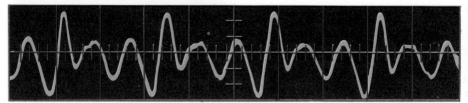

There are a great number of components in a violin note. Despite its jagged appearance on the oscilloscope the note does not sound jerky because the variations are so small that the ear does not pick them up individually.

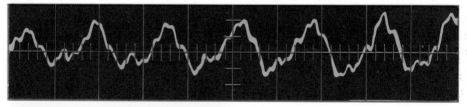

The screen records a split second in the life of one drum-beat. The irregularity shows that the sound is not a clear musical note with a regular wave pattern, but is a random collection of sound waves whose frequency and intensity constantly change.

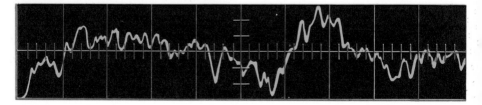

The instruments in an orchestra can be divided into three groups: those that produce sounds by striking (percussion) as in the drum, cymbal and triangle; those that have vibrating columns of air (wind instruments) as in the trumpet and flute, and those with vibrating strings (string section) such as the violin.

One of the simplest percussion instruments is the xylophone in which a series of wooden blocks of different lengths produce the different notes. The short blocks give the high notes and the long blocks the low notes. Under the blocks are tubes (resonators) of different lengths to intensify the sound. When the stretched skin of a drum is struck it vibrates and causes the air inside the drum to vibrate. In this way the vibrations are carried to the skin on the other side of the drum which also vibrates. The drum is tuned by altering the tightness of the skin.

In the brass instruments such as the trumpet, bugle and trombone the air is set vibrating by the player's lips. The players of woodwind instruments such as the bassoon, oboe and clarinet blow against the sharp edge of a flexible reed in the mouthpiece. The reed vibrates, which makes the column of air in the instrument vibrate in sympathy. The different notes are made by using the fingers to operate valves that cover and uncover holes in the body of the instrument. The length of the column of air depends on which holes are covered. In the recorder, which is the simplest wind instrument of all and has no valves, the holes are covered and uncovered with the fingers. The flute player blows across a small hole in the side of the instrument.

All wind instruments must have some way of changing the length of the vibrating air column to produce the different notes. In the slide trombone the length of the tube is changed. In other cases pistons and valves are used to make the air column longer or shorter.

The largest wind instrument is the organ which consists of a series of pipes of different lengths. Air is pumped into the pipes, and the keyboards and foot pedals control which pipes receive the air. The air vibrates in the pipes, which creates low notes in the long pipes and high notes in the short ones. The electronic organ is not a wind instrument because the notes are produced by electronic circuits.

Sound

97   *stringed instruments*

The best-known stringed instrument is the piano. When a key on the keyboard is pressed, a felt-covered hammer strikes three tightly-stretched metal strings, all tuned to the same note. Some upright pianos have only two strings. When it is not being sounded each string is prevented from vibrating by a felt pad (damper). When the key is pressed a system of levers first lifts the damper away from the string and then moves the hammer to hit the string. When the 'loud' pedal is pressed all the dampers are lifted leaving the strings free to vibrate. In some pianos the 'soft' pedal moves all the hammers nearer to the strings so that they do not strike so hard; in others it moves the whole keyboard and action (the levers and hammers) sideways so that only two of the three strings are struck.

Violin strings are made to vibrate by drawing a horsehair bow across them. The horsehair is first rubbed with rosin to make it slightly sticky. The string sticks to the bow, is drawn to one side and then springs back to the other side, and so on. This method of producing vibrations is called stick-and-slip action. The strings can also be set in motion by plucking (pizzicato). The sounds made by bowing

violin

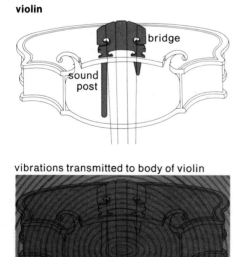

vibrations transmitted to body of violin

pedal harp

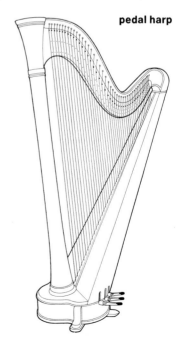

and plucking a string are different because of the different harmonics produced. The vibration of the strings is transmitted to the wooden body of the violin by the bridge which supports the strings. Under the bridge is the sound post which carries the vibrations from the belly of the violin body to the back. The whole body then vibrates in sympathy with the strings and causes large air vibrations.

The frequency with which a string vibrates can be changed by altering its length, its tightness (tension) or its mass per unit length. The four strings of a violin each have a different mass per unit length, the base G-string being the heaviest and the treble E-string the lightest. When the violinist tunes his instrument he turns the pegs and alters the tension in the strings so that they are tuned to the notes G, D, A and E. When he plays a note he either plays the open string or stops the string with his finger and thus changes its effective length.

The harp has 45 strings of fixed length and thickness. One end of each string is fixed to the sounding box and the other is attached to a metal tuning peg. The harpist can shorten the vibrating string with a foot pedal to raise the pitch of the note by a semitone.

grand piano

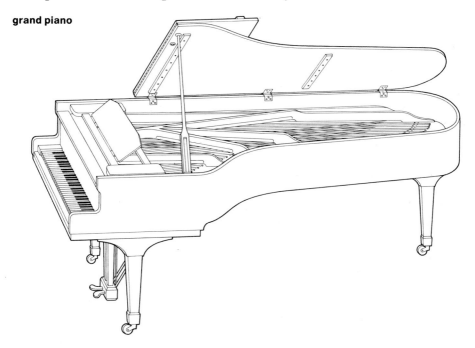

A vibrating object compresses the air next to it when it moves one way and allows the air to expand when it moves in the opposite direction. These compressions and rarefactions (expansions) are passed on to the neighbouring air molecules and spread out from the vibrating object as a sound wave. The air molecules move backwards and forwards in the same direction as the waves are travelling, forming a longitudinal wave. This is a different kind of wave from a light wave and radio wave – in these, the displacement is at right angles to the direction of travel and forms a transverse wave. A wave created by dropping a pebble into a pond is a transverse wave: the water moves up and down at right angles to the outward movement of the wave.

If the sound source is vibrating with a regular frequency the compression-rarefaction waves are produced regularly. The distance from one compression to the next is the wavelength. The speed of sound waves in air at sea-level and at 15°C is approximately 330 metres per second. The velocity (C), frequency (f) and wavelength (λ) are connected by a very simple relation: $C = f\lambda$.

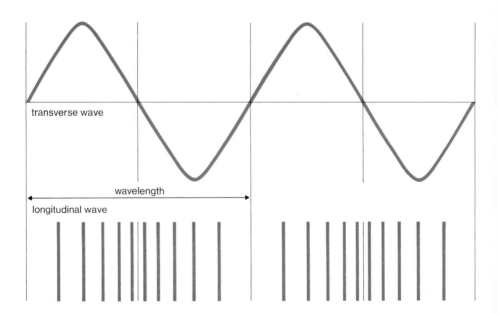

transverse wave

wavelength

longitudinal wave

When an express train sounding its whistle passes through a station, a person on the platform notices a sudden drop in pitch as the train passes him. The pitch of the sound is higher when approaching, and lower as it travels away. This change in pitch is known as the Doppler effect, named after the Austrian, Christian Doppler. In front of the moving source the waves are crowded together (shorter wavelength = higher frequency) while behind the moving source the waves are stretched out (longer wavelength = lower frequency).

Some modern aircraft can fly at speeds greater than the speed of sound: these are called supersonic aircraft. An object travelling at the speed of sound has a velocity of Mach 1. At twice the speed of sound its velocity is Mach 2, and so on. An approaching aircraft flying at Mach 1 produces sound waves that are crowded together at the same point. This creates a shock wave which causes a sonic boom when the plane crosses the sound barrier. The shock wave forms a cone with the nose of the plane at its tip: a boom carpet is swept out as this cone intersects the ground. At every point along this carpet the boom can be heard.

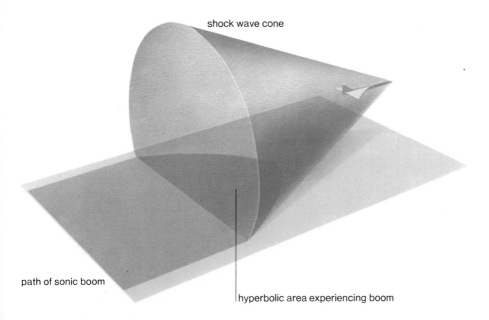

shock wave cone

path of sonic boom

hyperbolic area experiencing boom

Sound waves need a material substance in which to travel. Sound will not travel through a vacuum. This can be demonstrated by ringing an electric bell in a jar from which all the air has been gradually pumped out. When all the air has been extracted the sound from the bell can no longer be heard.

Sound travels better through solids than liquids, and better through liquids than gases. Many sounds are transmitted through the floors and walls of a building. But if you are in one room separated from another by a wall, not all of the sound will be transmitted through the wall from one room to the other. Some will be absorbed by the wall, which causes a tiny increase in the temperature of the wall: some will be reflected by the surface of the wall back into the room in which the sound occurs. You can hear the way sound is reflected if you walk down a narrow alley; your footsteps ring out as the sound is reflected backwards and forwards from the walls on either side of the alley.

Keeping sounds under control is one of the serious problems of modern life and sound-proofing a house can be just as important as

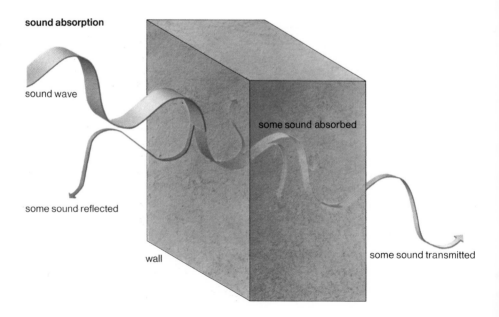

sound absorption

sound wave

some sound absorbed

some sound reflected

wall

some sound transmitted

making it heat- and draught-proof. However, what helps for one usually helps for the others. Keeping doors and windows closed, fitting double windows, building cavity walls and filling the space with expanded plastic or glass wool, all help to keep out unwanted sounds and to keep in the heat. Factory noise can sometimes be reduced by mounting machinery on sound-absorbing elastic pads or on springs and using polythene pipes in place of metal pipes.

Good acoustics in concert halls and theatres should enable the sounds from the stage to be transmitted to the auditorium so that listeners can hear them clearly and without distortion. Echo-reflections from hard-polished surfaces in or around the auditorium can distort the sound so that it may reach the listener by two or three different routes: as each route will take a slightly different time the sound will appear to be blurred. Sound-absorbent materials are used to cut out unwanted echoes. For example, in the Royal Festival Hall in London the tipped-up empty seats absorb approximately the same amount of sound as a person sitting in them. The hall, therefore, has good acoustics even when many of the seats are empty.

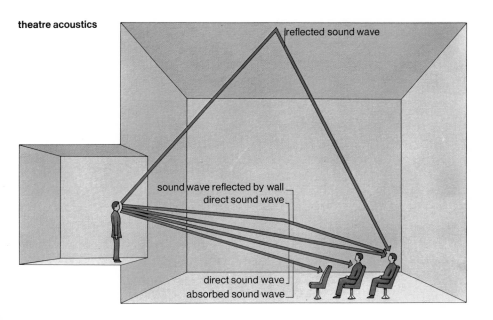

theatre acoustics

reflected sound wave

sound wave reflected by wall
direct sound wave

direct sound wave
absorbed sound wave

Sound
100 the ear and the telephone

The ear picks up sound waves in the air and converts them into nerve impulses which travel to the brain, producing the sensation of sound. The ear-drum is a paper-thin membrane which is set vibrating by the sound waves. This movement is carried by three small bones across the air-filled middle ear to the oval window of the inner ear. The middle ear is connected to the back of the throat by the Eustachian tube, keeping the pressure on both sides of the ear-drum the same.

The movements of the oval window in the inner ear cause pressure changes in the fluid-filled cochlea which is the real hearing organ. Inside the cochlea is a tapering membrane in contact with some 25 000 hairs of different lengths which are linked to nerve fibres. Sounds of different frequencies are detected by different sets of hairs. The ear is a very sensitive instrument and is capable of distinguishing between the frequency of sounds in the whole audible range. Our two ears acting together enable us to judge the direction of a sound. This is called the binaural effect.

The ear is not only a hearing organ but plays an important part in the balance of the body. Within the ear there are three semicircular

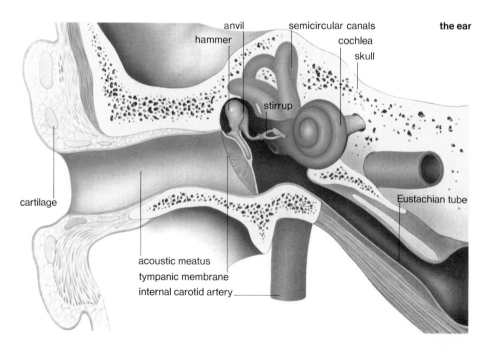

the ear

canals at right angles to each other which act like three spirit levels, to detect movements of the body. Spinning around quickly upsets the liquid in the balancing canals and causes a giddy feeling. There is also in the ear a gravity detector which tells us when we are upside down.

The ear converts pressure waves in the air into electric nerve impulses which are sent to the brain. We can think of the telephone as an extension of our ears. Pressure waves striking the diaphragm of the transmitter (microphone) at one end of a telephone link are converted into oscillating electric currents by the microphone. These currents can then be transmitted over any distance by telephone wires, radio links or by satellite. These electrical vibrations are converted back to pressure waves by the diaphragm in the receiver at the other end of the link. The receiver diaphragm vibrates with the same frequency as the transmitter diaphragm so that exactly the same sounds are heard. A battery or other source of electricity is included in the telephone circuit to provide power, and for long-distance calls the signals are amplified, or strengthened. To cross oceans, undersea cables, radio links or artificial satellites are used.

Sound
101 recording machines

In 1877 Thomas Alva Edison, the American inventor, devised a way to make sound vibrations produce a wavy groove on tinfoil wrapped around a revolving cylinder. A needle placed in the groove reproduced the vibrations on a thin diaphragm to which it was attached. The diaphragm was placed at the narrow end of a horn which sent out the sound waves. The tinfoil on the recording cylinder was later replaced by wax, which in turn gave way to a flat, shellac disc and finally to the plastic (vinyl) record. In early machines the cutting of the groove in the record was done mechanically. Modern techniques of sound recording are electrical.

The sound waves are picked up by a microphone and converted into oscillating electric currents. The currents are made to vibrate a diamond point (stylus) in tune with the sound. The vibrating stylus cuts a groove in a rotating lacquered aluminium disc to make the master copy. The master is then nickel-plated, and when the plating is stripped off, a disc with ridges instead of grooves is formed. By plating and stripping again, an exact copy of the master is made, which is known as the mother. From the mother new ridged plates,

gramophone 1907

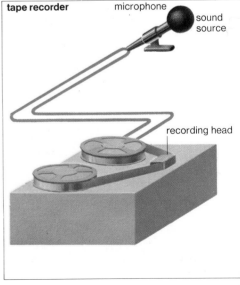

tape recorder   microphone   sound source

recording head

the stampers, are made which are used to press out the final records.

Since about 1950 the first stage in making a record has been to produce a magnetic tape of the sounds to be recorded. A narrow plastic tape coated with powdered iron oxide runs past a gap in an electromagnet, the coil of which is connected to the recording microphone. The sound waves are converted by the microphone into a current, which alters the strength of the electromagnet and changes the pattern of magnetization of the powder particles on the tape. The tape is run past the gap again, when the coil of the electromagnet is connected to an amplifier and loudspeaker, thus reproducing the sound. This method of recording by tape reproduces more faithfully the received sounds and has led to the high-fidelity (hi-fi) recording systems. In addition to making records, these tapes can be copied and are sold in cassettes for playing directly on tape recorders.

For stereophonic records the sound from the left-hand side of the orchestra is recorded on the inner side of the record grooves while sound from the right-hand side of the orchestra is on the outer part of the groove. Stereophonic cassettes are also widely used.

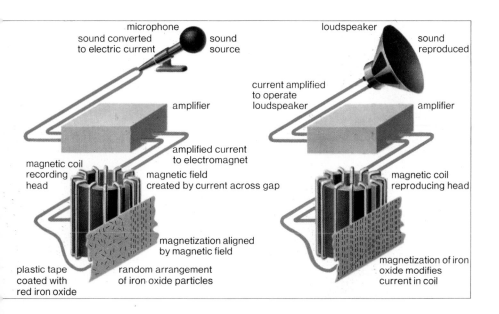

The human ear will respond to frequencies from about 20–20 000 hertz (cycles per second). Lower frequencies are called infrasonic and higher frequencies are called ultrasonic. Different animals have their own particular audible ranges. Some insects are sensitive to frequencies up to 40 000 hertz: bats navigate through dark caves by sending out ultrasonic bleeps which echo back to them like radar signals. Dogs can hear higher frequencies than humans so a dog can be called by using a Galton's whistle, which gives out very high-pitched notes which the dog can hear but which humans cannot. Ultrasonic waves may have frequencies up to several million hertz and therefore their wavelengths can be as short as 0·1 mm.

Metal diaphragms and loudspeakers are too heavy to vibrate at these frequencies, so that in order to produce short ultrasonic waves a different kind of transducer is needed. A transducer is a device that changes one kind of oscillation into another. The most common type makes use of a piezo-electric crystal. This is a special kind of crystal, such as quartz or Rochelle salt, that can be made to vibrate when an oscillating current is fed across its faces.

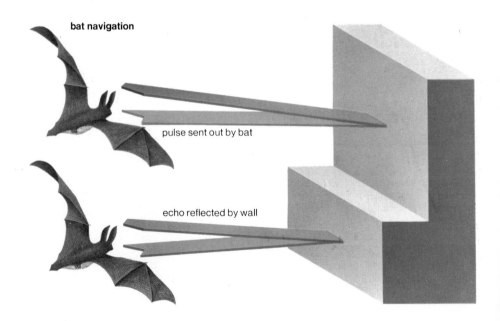

bat navigation

pulse sent out by bat

echo reflected by wall

In the same way as bats use sound echoes, ship navigators explore underwater with ultrasonic waves. Their sonar (*so*und *n*avigation *a*nd *r*anging) sends out pulses of ultrasonic vibrations, which are reflected back to the ship from the bottom of the ocean. By measuring the time taken for the echo to return it is possible to calculate the depth of the sea. This enables a map of the sea-bed to be drawn. The same principle is used to detect flaws in metal castings. In a solid piece of metal, reflection of the waves is usually from the bottom surface. But if there is a flaw in the metal, reflection will occur from the base of the flaw. Aluminium castings for aircraft parts are tested in this way to make sure that they will not fail in flight.

Ultrasonics can also be used for cleaning by shaking dirty clothes in water or other cleaning fluid. Ultrasonic agitation is used in cleaning delicate parts of machines, even watches.

The oscillations of an ultrasonic beam directed at small living organisms, such as bacteria, make the cell walls vibrate in sympathy. This breaks up the walls and destroys the organism. Ultrasonics can thus be used in food preservation and in medicine to kill germs.

sonar

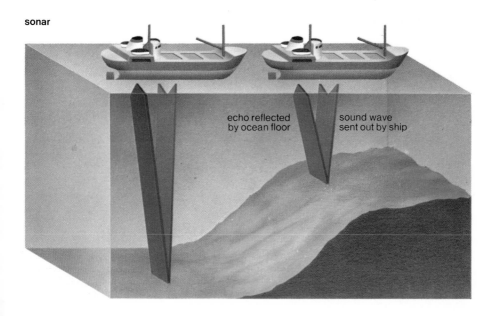

echo reflected
by ocean floor

sound wave
sent out by ship

When a perspex rod is rubbed with a cloth, the rod will pick up scraps of paper, in the same way that a magnet attracts iron filings. As long ago as 600 B.C., a Greek named Thales found that amber had this property. An English physician, William Gilbert, over 2000 years later coined the word electricity from the Greek word *elektron*, meaning amber. It was not until 1733 that a French scientist, Charles du Fay, discovered that there are two kinds of electricity. If two amber beads are electrified by being rubbed, and then suspended close together on separate cotton threads, they will repel each other. Two glass beads, after rubbing, behave in the same way. But if one glass bead and one amber bead are suspended next to each other they attract one another. The two kinds of electric charge are called positive and negative: similar charges (two positives or two negatives) repel each other, but dissimilar charges (one positive and one negative) attract one another.

These two types of electricity are caused by the movements of negatively charged electrons from one material to the other. The

Benjamin Franklin, an American scientist, demonstrates the electrical nature of lightning. As a result of this experiment in June 1752, he invented the lightning rod

atoms of an uncharged body contain equal numbers of electrons (negative) and protons (positive). They are therefore electrically neutral. Removal of electrons leaves a body positively charged, while an accumulation of extra electrons makes it negatively charged.

The kind of electricity in which charges build up on insulating materials is static electricity, as opposed to the moving current electricity that flows through conductors such as metals. It is also sometimes called frictional electricity because it is produced by rubbing. It is very difficult to imagine modern life without electric currents for lighting, heating and electric motors, but frictional electricity was the only kind known until the end of the 18th century.

Modern materials – plastics and synthetic fibres – easily become charged with static electricity which can be a nuisance. Constant rubbing on the plastic-covered seats of a car charges both the car and the passenger who may receive a shock on touching the metal door handle. Objects made of synthetic plastics charge easily in dry, air-conditioned buildings so a person receives an electric shock when he touches them.

Static electricity makes the balloon cling to the child's clothing

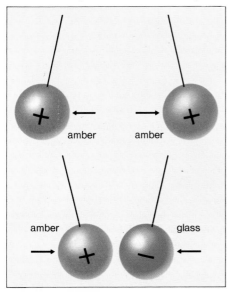

An electric current is a flow of electrons. Thus current electricity and static electricity are essentially the same, but one consists of moving charges and the other of stationary charges.

In 1791, Luigi Galvani, an Italian anatomy teacher, accidentally discovered that he could make a dead frog's leg twitch by touching it with two different metals. His fellow countryman Alessandro Volta, a physicist, believed that the electricity making the leg twitch was not in the muscle but had something to do with the junction of the two metals. He therefore tested various combinations of metals connected by salt solutions rather than muscle. From his experiments he produced the first electric battery and the first electric current. Volta's pile, as it was called, consisted of a series of zinc and copper discs separated by fabric discs soaked in salt solution. When a wire was connected to the top and bottom discs an electric current flowed.

Many different types of battery have been invented since Volta's pile but they are all made up of two different elements in a solution or paste of a chemical salt. Today, the type most commonly used in torches and radios is the dry cell. When one of the components of the

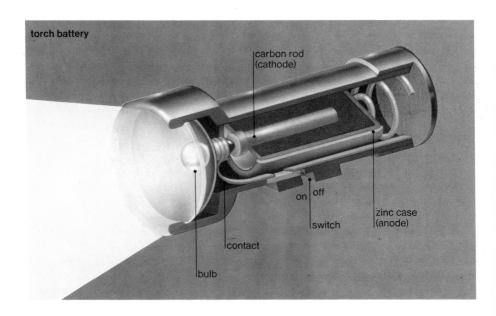

torch battery

carbon rod
(cathode)

on off

zinc case
(anode)

switch

contact

bulb

cell is used up the cell ceases to give a current. A cell which produces current from chemicals in this way is known as a primary cell. A cell has two terminals to which the connections are made. They are often marked positive ($+$) and negative ($-$). The electrons flow from the negative terminal around to the positive terminal.

A secondary cell, or accumulator, is one in which the chemical reactions are reversible. This cell can be recharged when it is run down by passing an electric current through it in the opposite direction to the flow of current when the battery is in use. The most common type of storage battery is the lead-acid accumulator which is used in most cars. The battery supplies the current for the starting motor and lights when the engine is not running and is recharged by the dynamo or alternator driven by the engine. The lead-acid accumulator consists of lead plates immersed in dilute sulphuric acid. When current is being taken from the battery the lead plates tend to become lead sulphate. When the accumulator is being charged the sulphate ions go back into solution. A type of accumulator called a Ni-fe cell has nickel and iron electrodes immersed in potassium hydroxide.

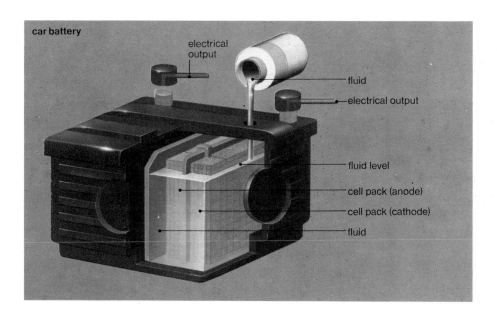

car battery

electrical output

fluid

electrical output

fluid level

cell pack (anode)

cell pack (cathode)

fluid

An electric current is a flow of electrons. The path along which the electrons flow is called a circuit. For an electric current to flow there must be a complete circuit without gaps. A simple circuit might consist of a battery, a bell push and an electric bell. When the button is pressed the circuit is completed and the bell rings as the current flows through the electromagnet. Releasing the button breaks the circuit and the bell stops ringing. This is a simple series circuit but, as the photograph of a complex electronic circuit shows, a circuit may be very intricate with a variety of parallel paths for the electron stream to follow.

Substances which allow an electric current to flow through them easily are called conductors. Metals are good conductors, solutions of some chemicals (acids, bases and salts) will also conduct, but not as well as metals. This is because metals contain many free electrons (electrons not attached to any particular atom) that can move when an electric potential is applied to them. Substances which will not allow a current to pass through them are insulators or non-conductors: rubber, china, plastics and most non-metallic substances.

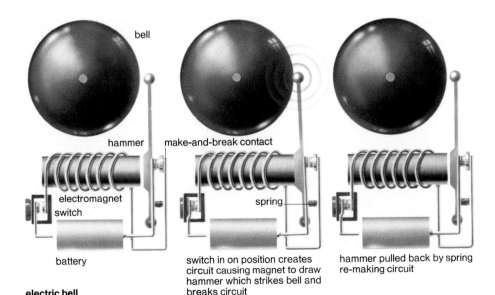

electric bell

switch in on position creates circuit causing magnet to draw hammer which strikes bell and breaks circuit

hammer pulled back by spring re-making circuit

Wires carrying electric current almost always have an insulating covering to prevent the current leaking away. If two bare wires touch a short circuit occurs causing a spark.

Semiconductors can conduct electricity better than insulators, but not as well as metals: germanium, silicon and selenium are semiconductors. In the pure state they have very few free electrons but the number can be increased by adding a small number of special impurity atoms. Transistors consist of different types of semiconductors joined together, often in the form of a sandwich.

An electric current is a flow of electric charge and a flow has direction. Before the discovery of electrons at the end of the 19th century, scientists decided to say that the current flowed from the positive terminal around the circuit to the negative terminal. This is still the direction shown on some circuit diagrams and is known as the conventional current. The real electron flow, called the electron current, is in the opposite direction. The negative terminal is the source from which electrons (which are negatively charged) enter a circuit: they leave by the positive terminal.

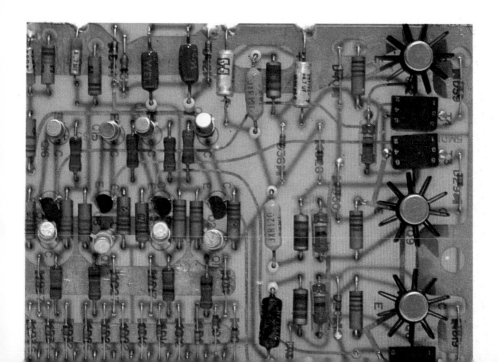

The flow of electricity through a wire is often compared with the flow of water through a pipe. Since the pipe offers a resistance to the flow, a pressure is needed to drive the water along. This pressure can be produced by a pump as it is in some central heating systems. Similarly, with an electrical circuit there has to be an electric pressure to cause the electrons to flow. This is provided by the battery or generator, which acts as a kind of electron pump.

The electrical pressure, or electromotive force, is measured in volts. The current, or rate of flow of electricity, is measured in amperes. One ampere is equivalent to a flow of about six million million million electrons in each second. The resistance of the conductor to the electron flow is measured in ohms. The names (volts, amperes, ohms) of these units commemorate famous scientists.

A very simple relationship exists between the three quantities:

$$\text{current} = \frac{\text{voltage}}{\text{resistance}} \quad \text{or} \quad \text{amperes} = \frac{\text{volts}}{\text{ohms}}$$

The relationship was discovered by Georg Ohm, a German physicist, in 1826 and is known as Ohm's Law. All simple electrical calculations are based on this law. If two of the three values are known the third can be calculated.

Electrical power (rate of working) is measured in watts or the larger unit the kilowatt ($= 1000$ watts). The power used depends on both current and voltage and, in fact, watts $=$ amperes $\times$ volts. This is another useful relation for calculations. For example, a 2 kW electric heater connected to a mains supply of 240V will use $\frac{2000}{240} = 8\cdot3$ amperes. This is why two such heaters must not be connected to a plug containing a 13 amperes fuse. The double current of 16·6 amperes will blow the fuse.

The electric meter records both the power and the time for which it is used. It therefore measures electrical energy, usually in units of kilowatt-hours; this is the amount of energy used when an appliance rated at 1 kW is used for one hour. A 100 W lamp running for 10 hours uses one unit of electricity: a 2 kW heater working for 10 hours uses 20 units. Our electricity bills tell us how many units we have consumed.

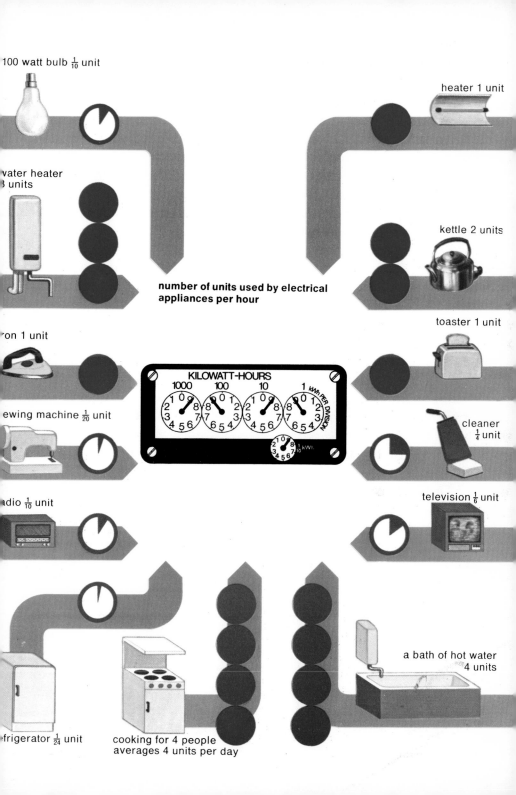

When an electric current flows through a conductor the atoms or molecules of the conductor offer a resistance to the flow of electrons. Some substances, such as silver and copper, offer very little resistance and are good conductors. Others, such as glass and polythene, offer a great deal of resistance and are bad conductors but excellent insulators. Resistance also depends on other factors: the longer and thinner a piece of wire, the greater will be its resistance. Its resistance also increases when it is heated.

When a current is forced through a conductor against its resistance, the electrons collide with the atoms of the conductor and make them move about faster. This causes the temperature of the conductor to rise. A practical use of this is in electric heaters. If the temperature rise is great enough light is produced and this is used in electric lamps.

The first electric lamps were made in 1860 by Sir Joseph Swan in England, and developed commercially by Thomas Edison in the United States some twenty years later. They used a filament made of bamboo fibre covered with carbon; this was enclosed in a glass bulb from which all the air had been pumped to prevent the filament from

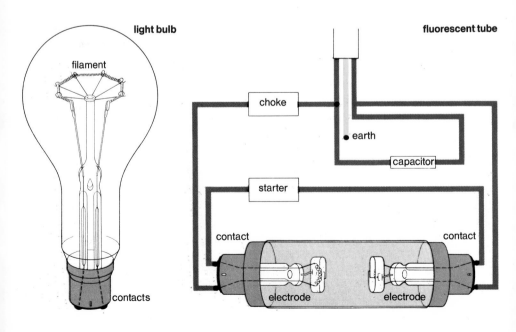

light bulb

filament

contacts

fluorescent tube

choke

earth

capacitor

starter

contact

contact

electrode

electrode

burning away. Modern lamps have coiled filaments made of tungsten which become white hot (incandescent) at 1300°C. The bulbs are filled with inert gases such as argon or nitrogen at low pressure. Although they give more light for less electric power than incandescent lamps they still waste a great deal of energy as heat.

Fluorescent tubes produce much less heat and are therefore much more efficient. Electrodes at each end of the tube are heated so that they give off electrons which, instead of flowing through a metal filament, flow through mercury vapour. When the electrons collide with the mercury atoms these atoms become excited (raised to a higher energy level); when they return to the ground state – the original state before they were excited – they give off ultraviolet radiation. This strikes the chemical coating (phosphor) on the inside of the tube making it glow with a soft, bluish shadow-free light.

Discharge lamps, in which a stream of electrons pass through a gas, have replaced the old gas lamps for street lighting and are widely used for advertising signs. Neon gas gives a red light, nitrogen a pale yellow, sodium an orange, and mercury vapour a blue light.

fluorescent tube

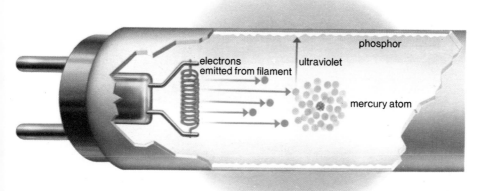

The ancient Greeks knew about magnets – they knew that pieces of iron ore, magnetite or lodestone, have the property of attracting fragments of iron. By the beginning of the 12th century the Chinese had discovered that if a piece of lodestone was suspended it always set itself so that it pointed approximately north and south. By the end of the same century European ships were using this simple compass as an aid to navigation.

In the 16th century William Gilbert, Queen Elizabeth's physician, found that the properties of the lodestone could be transferred to an ordinary piece of iron by rubbing it with the lodestone. By this method he made an artificial magnet from a natural one. He also suggested that a compass always points north and south because the Earth itself has magnetic properties.

The power of a magnet to attract is greatest at its ends, which are called the poles. The pole that points towards the north is the north pole, the other the south pole. The space around a magnet, in which it exerts its attractive or repulsive force, is its magnetic field. Experiments with magnets show that unlike poles attract each other, but

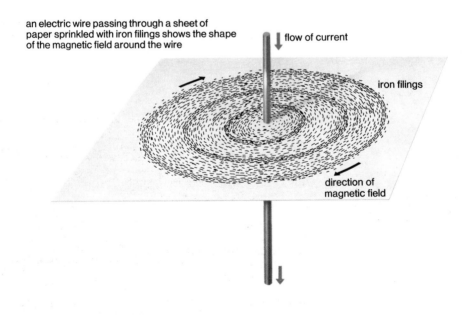

an electric wire passing through a sheet of paper sprinkled with iron filings shows the shape of the magnetic field around the wire

flow of current

iron filings

direction of magnetic field

similar poles (two north poles or two south poles) repel each other. In fact, poles behave very much like electric charges except that poles always exist in pairs, never alone. No one has ever found a magnetic monopole.

This connection between magnetism and electricity was first investigated in 1819 by the Danish scientist, Hans Oersted. He found that a wire carrying an electric current affected the position of a compass needle placed close to it. He concluded that an electric current has a magnetic effect, and that a conductor carrying a current is surrounded by a magnetic field. This discovery was used by an English scientist, William Sturgeon, in 1825 to make an electromagnet, which consists of a soft iron bar with a coil of wire around it. When the current in the coil is switched on the bar becomes magnetized; when it is off the bar loses its magnetism.

The ability to turn a magnet on and off makes the electromagnet a very useful device. Huge magnets carry and lift scrap iron. Smaller ones form essential components of electric bells, telephones, electric motors and dynamos.

electromagnet

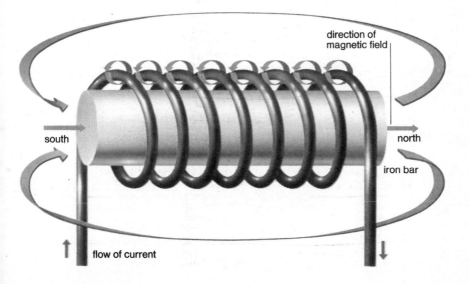

direction of magnetic field

south

north

iron bar

flow of current

how a generator works

a coil is turned in a clockwise direction between opposite poles of two magnets

magnet   coil | |magnetic field   magnet

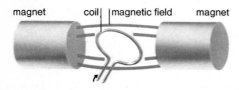

coil interrupts lines of magnetic field causing flow of electric current through coil

coil interrupts maximum number of lines creating peak flow of current

magnetic field uninterrupted so no current flows

coil turned through half revolution reversing flow

coil turned through full revolution so current flows again in original direction

One of the most important scientific discoveries was made by Michael Faraday in 1831 at the laboratories of the Royal Institution in London. Faraday found that when a magnet is moved near a coil of wire, an electric current is produced in the coil. He also discovered that when a current is switched on and off in one coil a changing magnetic field is created around the coil and that this changing field will create another current in a second nearby coil. The current in the second coil is much greater if both coils are wound on the same iron core. This process of producing electricity from magnetism is called electromagnetic induction.

The electric generators, or dynamos, that supply us with the vast amounts of electric power needed today have all been developed from Faraday's discovery. A dynamo consists of many coils of wire wound on an iron core (armature) which is rotated on a shaft between the poles of an electromagnet. The current generated in the coils when the armature is driven around is led away by sliding carbon contacts (brushes) which rub against slip rings at the end of the shaft. In a simple dynamo the current changes direction twice for each revolution of the coil. First it builds up to a positive peak, then it reverses and falls to a negative peak. It then returns to zero before building up to a positive peak again. This is called alternating current, and a graph showing how the current varies with time has the shape of a wave. To obtain direct current from a rotating generator requires a commutator with split rings.

Generators do not make electricity from nothing. They have to be driven by some outside source of energy such as a steam turbine. The steam is made by burning coal or oil or by using nuclear energy, and the turbine turns the generator, thus converting mechanical energy into electrical energy. In a hydroelectric power station the generator is driven by a water turbine.

Once the principle of electromagnetic induction had been discovered there was rapid development in the design of more efficient generators. By 1882 the streets of New York were lit by over two thousand electric lamps. Today cities and most industries are dependent on power stations and their generators.

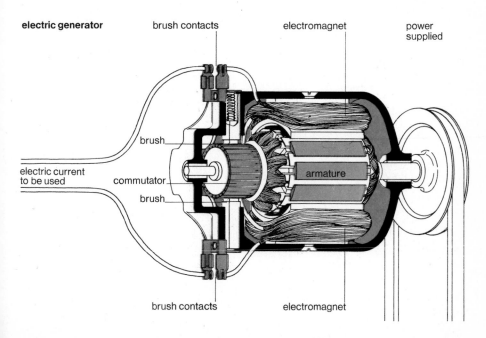

electric generator brush contacts electromagnet power supplied

brush

electric current to be used commutator armature

brush

brush contacts electromagnet

An electric motor is the reverse of a generator – it is a machine for converting electrical energy into mechanical energy. Electric motors are extremely efficient and because they are quiet and clean, can be easily switched on and off and are made in a wide variety of sizes, they have many uses.

When a conductor carrying an electric current is placed in a magnetic field a force is produced on the conductor. Also, when the magnetic fields of two electromagnets overlap, the two electromagnets are forced apart. These are the principles of the motor. In a motor the current is fed in by the brushes to the rotating coil (rotor), instead of being taken out by them as they are in a generator. The rotor thus becomes an electromagnet. Around this coil is another coil (the field coil or stator) which is stationary and must also be supplied with current. The force between the magnetic fields of the rotor and the stator makes the rotor spin. Motors can be designed to work on direct or alternating current.

A synchronous motor will only work on alternating current and revolves at a rate controlled by the frequency of the supply. Since the

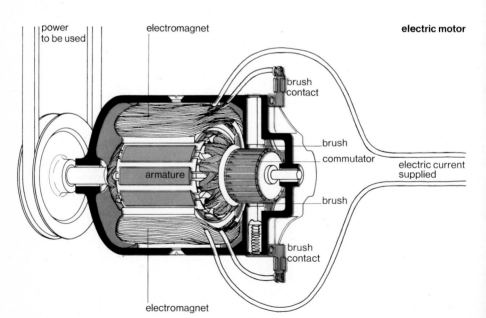

power to be used · electromagnet · **electric motor** · brush contact · brush · commutator · electric current supplied · armature · brush · brush contact · electromagnet

mains frequency (50 hertz) is kept constant at power stations, synchronous motors are used to drive electric clocks. But any variation in the frequency will make the clock run too slow or too fast.

Another special type of alternating current motor is the induction motor. Current is fed to the stator only, the rotor current being induced by the stator current. A great advantage of the induction motor is that the current does not have to be led into the rotating coils through brushes as in an ordinary electric motor. Most large motors have rotors with several separate coils set at an angle to each other. In these motors, the commutator is split into segments each of which is connected to one of the coils and is fed with current by the brushes.

The linear induction motor is a new development which produces movement in a line instead of a rotation. The stator coils are arranged on either side of an aluminium or copper track. The changing fields of the stator coils induce currents in the track and the stator moves forwards. This type of motor has driven experimental monorail trains.

how an electric motor works

a current is passed through the coil causing it to turn in a clockwise direction

magnet   coil   magnetic field   magnet

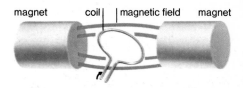

the interaction between the magnetic field and the current causes the coil to turn

when the maximum number of lines of the field are cut by the coil the torque is greatest

when no lines are being cut there is no torque the motor would not start from this position

In a wire carrying a direct current (produced, for instance, by a battery) there is a constant stream of electrons in one direction along the wire. An alternating current reverses its direction at regular intervals. In a wire carrying an alternating current the electrons surge rhythmically backwards and forwards. One complete forward and backward movement is a cycle and the number of cycles per second is the frequency of the alternating current. This is measured in hertz (one hertz = one cycle per second). The frequency supplied to homes in Britain is 50 hertz, and in the United States it is 60 hertz.

An alternating current operates lights, heaters and motors just as well as a direct current but it has the great advantage that the voltage of the supply can be easily changed up or down by using a transformer. A transformer consists essentially of two separate coils of wire wound onto the same soft iron core. Depending on the relative number of turns on the primary (input) and secondary (output) coils the voltage can be either stepped up or stepped down. If there are 100 turns on the primary coil and 200 turns on the secondary coil, the voltage will be doubled. But the power remains the same, apart

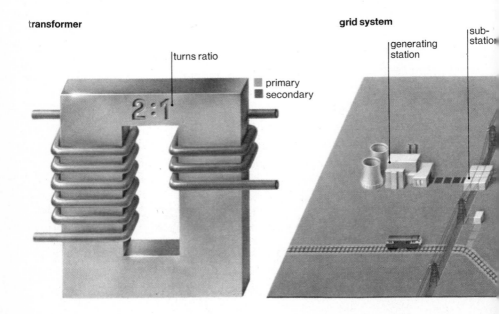

transformer

turns ratio

■ primary
■ secondary

2:1

grid system

generating station

sub-station

from heat losses, so the secondary current will, in this case, be half that of the primary. As a current is only induced in a secondary coil when the field around the primary coil is changing, a transformer will not work on direct current.

Large transformers are used in the grid system which links all the power stations in Great Britain so that current from one station can be supplied to another region in the case of a breakdown or a greater demand for current than usual. The large power station generators produce electricity at 6000 V. However, it is cheaper to transmit a low current at a high voltage, rather than a high current at a low voltage. This is because the power loss is proportional to the square of the current, but only directly proportional to the voltage. Also a thinner (and therefore cheaper) cable can be used at low currents. The output of the power station is thus stepped up to 132 000 V (132 kilovolts) before being fed into the grid system. It is stepped down again at local sub-stations and reaches the consumer at 240 V. Some modern super-grid systems have a voltage of 400 000 V (400 kV) and in the United States voltages over 700 000 V (700 kV) are in use.

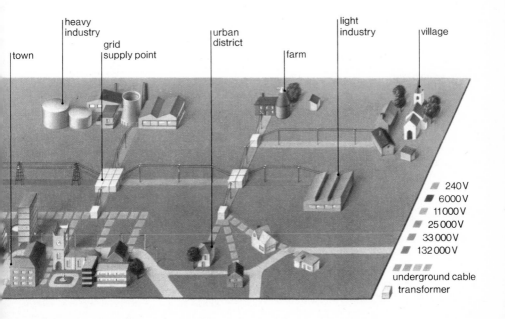

town

heavy industry

grid supply point

urban district

farm

light industry

village

- ▨ 240 V
- ◼ 6000 V
- ▨ 11 000 V
- ▨ 25 000 V
- ▨ 33 000 V
- ▨ 132 000 V

▨▨▨▨
underground cable
▱ transformer

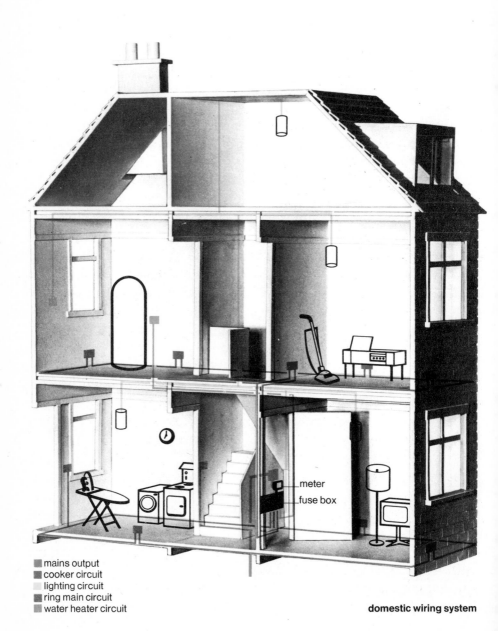

meter

fuse box

mains output
cooker circuit
lighting circuit
ring main circuit
water heater circuit

domestic wiring system

A modern power station generates enough electricity to supply a large area. The current can be easily transmitted by cables which makes electricity the simplest means of supplying homes and factories with a source of energy. It is used for lighting, heating and driving electric motors, both for the large motors used in industry and the small ones used for domestic appliances.

Electricity enters a building through a fuse box, meter and main switch. A fuse is a thin piece of wire made from an alloy of tin and lead with a low melting point. If, due to a short circuit or for any other reason, a larger current than is marked on the fuse flows through it, the fuse heats up quickly and melts, thus breaking the circuit and preventing damage to any appliances. It is also an essential fire prevention device: wires with faulty insulation could cause sparks if they were permitted to short circuit. The main switch connects the house circuits to the main electricity supply: when it is off, all circuits are disconnected and are safe to handle.

Modern houses are wired with a ring circuit. Older houses have several outlets connected to the same fuse so that a fault in one appliance blows the fuse and puts all the outlets out of action. In the ring system each plug has its own fuse. If this blows it can be easily replaced and does not affect other equipment. The ring system is cheaper to install because it uses less wire than the older method. Each ring circuit can supply a current of about thirty amperes so larger houses may be wired with two or more separate ring circuits. If there is an electric cooker it is wired separately because it needs a considerable amount, up to forty-five amperes, of current.

One wire in the system is live and the other is neutral. A third contact in each socket is connected directly to Earth. Appliances such as kettles, irons and radiators are connected with these three wires. If the live wire accidentally comes in contact with the metal part of an appliance the current will flow directly to Earth and cannot cause a shock to anyone handling it. Electric devices should never be touched with wet hands. Damp skin has a much lower resistance than dry skin, which is why a dangerously high current flows through the body if there is an accidental leak.

Electrons, the tiny particles of negative electricity that revolve around the nuclei of atoms, were discovered by the English physicist, J. J. Thomson, in 1897. It then appeared to be an isolated discovery but in the following 75 years it has led to the growth of a vast industry that has changed our lives in many ways. At first, all of these industries relied on the thermionic valve, invented in England by John Ambrose Fleming in 1904. The valve was the first electronic device to be invented, and its development revolutionized methods of communication and entertainment.

In 1948 the transistor was invented by William Shockley, an American physicist. It depends on the behaviour of electrons in semiconductors such as silicon and germanium. Transistors can do all that valves can do but they are much smaller, and therefore lighter, stronger, and work with much smaller power supplies than valves. This led to much smaller, or miniaturized, electronic equipment. The electronic amplifier for a hearing-aid can be made small enough to be concealed behind the ear. Radio and television equipment can now fit in the restricted space of a satellite. In fact, the exploration of

John Ambrose Fleming, an English physicist, invented the thermionic valve

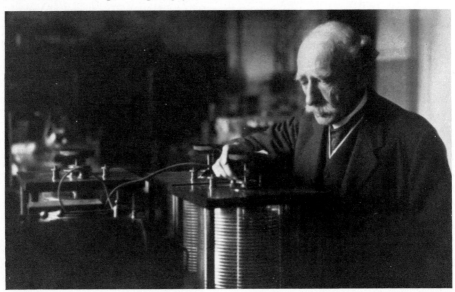

space would not have been possible without transistors. Integrated circuits – the smallest electronic devices – are made by designing semiconducting crystals to perform the functions of whole circuits. In these circuits, a chip of silicon may contain a large number of separate components, such as transistors, connected together.

Computers also depend on electronic switches. The speed with which the electronic switches work allows the computer to carry out very complicated mathematical operations very quickly. In a few seconds, long and difficult problems can be solved which would take mathematicians years to work out by ordinary methods. But the computer must be programmed with great precision by the mathematician so that each step follows logically from the one that precedes it. Computers play an essential part in the automation of complex manufacturing methods. They monitor the processes and feed instructions into the control equipment. They also have a wide variety of other uses which range from keeping the accounts in large stores, to mapping out the course of a moon rocket.

Left: *J. J. Thomson, by his discovery of electrons, revolutionized modern physics.*
Right: *some of the micro-miniaturized circuit elements used in computers*

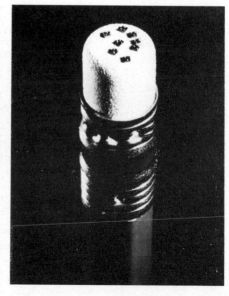

The transmission of messages through space without wires connecting the sender and receiver, as they do in the telegraph and telephone, grew out of electromagnetic theories put forward by the Scottish physicist, James Clerk-Maxwell, in about 1865. Some twenty years later Heinrich Hertz, in Germany, detected and produced electromagnetic (radio) waves for the first time. He held a ring with a tiny gap in it a few yards away from a sparking coil and discovered that sparks were produced in the gap. The electromagnetic energy had been transmitted through space from the coil to the ring.

The whole science of radio communication stemmed from this one laboratory experiment. New methods of detecting radio waves were invented in England, France and Russia, but the first practical use of the discovery was made by an Italian, Guglielmo Marconi. In 1901 he sent a message nearly 3500 km across the Atlantic Ocean.

These early messages were sent in Morse code, a system of short and long signals invented by an American, Samuel F. B. Morse. Each letter of the alphabet is represented by its own combination of dots and dashes. The radio waves were sent out in groups of short and

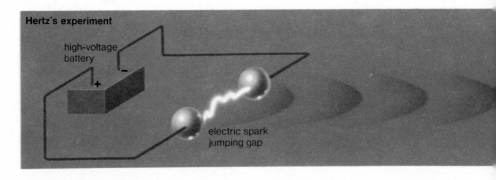

Hertz's experiment

high-voltage battery

electric spark jumping gap

international Morse code

A B C D E F G H I J K L M

long signals which operated a buzzer at the receiving end. The buzzes had to be decoded before the message could be read.

Wireless waves could not carry speech until a method had been developed for combining the low-frequency (audio-frequency) currents produced in a microphone by sound waves with the much higher-frequency currents that produce radio waves. This only became possible after the invention of the thermionic valve. The valve can be used as an oscillator to produce the high-frequency currents that produce radio waves: these high-frequency currents are then modulated (increased and decreased in amplitude) by the currents produced by sound waves in the microphone. The modulated current is fed to the transmitting aerial and the radio waves are broadcast into space. They are received by the receiving aerial and the pattern of the sound wave is separated from the radio wave by another valve (detector). The audio-frequency current is then amplified to increase its strength, and finally the original sound waves are reproduced by a loudspeaker or earphones. Thermionic valves have now been superseded to a great extent by transistors.

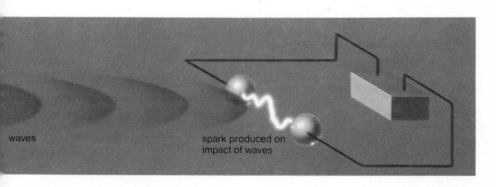

waves

spark produced on impact of waves

N O P Q R S T U V W X Y Z

Metals contain electrons which move about freely rather than being attached to atoms. When a metal is heated in a vacuum these electrons move about more energetically and some will have sufficient energy to escape from the surface of the hot metal. This is thermionic emission, which is somewhat similar to the evaporation of molecules from the surface of a liquid.

In an ordinary electric lamp the electrons emitted by the filament form a cloud around it. If a metal plate is sealed inside the lamp bulb, a little distance from the filament, a simple diode valve is formed. If the positive terminal of a battery is connected to the plate (anode) and the negative terminal is connected to the filament (cathode), the negatively charged electrons will no longer collect around the filament but will be attracted towards the anode. Because a flow of electrons constitutes an electric current, a current flows through the valve from the cathode to the anode. If the anode is made negative it will repel rather than attract electrons and no current can pass through the valve. Thus the valve acts as a one-way flow regulator for electrons very much like a valve in a water pipe. A valve operating

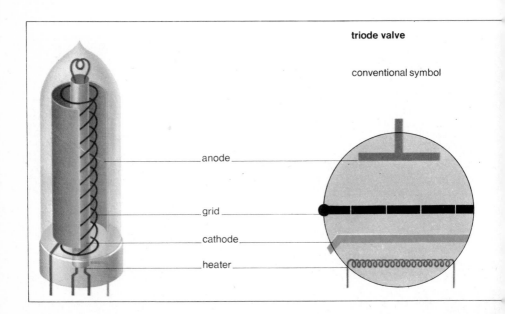

triode valve

conventional symbol

anode

grid

cathode

heater

in this way can be used as a rectifier: by cutting off the negative swing of an alternating current it converts it into direct current.

An American inventor, Lee de Forest, added a third electrode in the form of a wire spiral between the cathode and the anode and produced the triode valve. This third electrode (grid) can control the number of electrons reaching the anode. The more positive the voltage of the grid compared to the cathode, the more electrons pass through it to reach the anode and so the anode current is increased. The less positive the grid, the fewer the number of electrons to reach the anode. The important point is that a small variation in the voltage of the grid can produce a large variation in the anode current. This is rather like a stick pivoted at one end. A small movement of the stick halfway along its length will cause a much larger movement at the free end. So if a weak signal is fed to the grid a much stronger signal can be taken from the anode. The valve therefore acts as an amplifier. Modern valves have several grids and are more sensitive than the triode. Until the invention of the transistor, all radio and record-player amplifiers were based on thermionic valves.

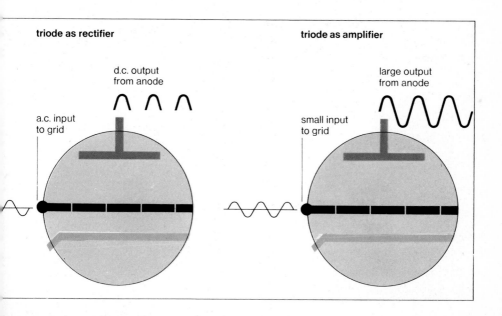

triode as rectifier

triode as amplifier

d.c. output
from anode

large output
from anode

a.c. input
to grid

small input
to grid

Valves are used in electronics to control the flow of currents in circuits, to detect and amplify radio signals and to produce oscillations in transmitters. Transistors can also perform these functions. In valves, the current is carried by electrons passing from the cathode to the anode in a glass tube from which the air has been removed. Transistors are completely solid, the electrons moving between the atoms of certain semiconducting crystals. For this reason they are called solid-state devices.

Pure germanium, which is a typical semiconductor, has very few free electrons to carry a current. The four outer electrons of the germanium atom are able to form links (covalent bonds) with electrons in adjacent germanium atoms. However, if an atom with five outer electrons, such as arsenic, is introduced, four of the electrons link with four germanium atoms leaving one extra, free electron to act as a current carrier. This is known as *n*-type germanium because the current carriers are *n*egative electrons; *n*-type germanium is a better conductor than pure germanium because it contains more free electrons. If an atom with three outer electrons, for example alu-

William Shockley and his co-workers, the inventors of the transistor

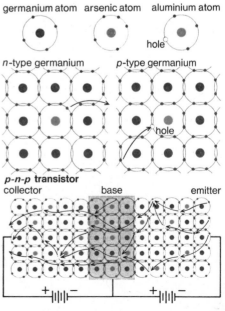

minium or indium, is added to germanium, an electron gap or hole is formed. This hole can be filled by an electron from a neighbouring atom leaving another hole in its place. As the electrons (negative) move in one direction, the holes (positive) move in the opposite direction. This is *p*-type germanium in which the holes behave like *positive* electrons (positrons).

A transistor is made by sandwiching a thin layer of *n*-type germanium between two small pieces of *p*-type germanium. This is a *p-n-p* transistor (*n-p-n* transistors are also used). The central *n*-slice is the base and corresponds to the grid of a valve. One of the two outer *p*-type layers is the emitter and the other is the collector: they correspond to the cathode and anode of a valve, respectively.

Transistors are very small and light. A typical transistor is less than one centimetre long and three millimetres in diameter. There is no filament to heat up in order to emit electrons and so there is no warming-up time. Modern radio and television sets and computers use transistors instead of valves.

Comparison of the sizes of a valve (left), *a transistor within a protective can* (centre) *and a silicon integrated circuit containing 22 transistors* (right)

The first radio messages were sent in Morse code. The radio waves from the transmitter are switched on and off by pressing and releasing a tapping key to form a pattern of dots and dashes which are then picked up by the aerial of the receiver.

For the transmission of speech and music the sound waves are converted by a microphone into an oscillating electric current which varies at the same frequency (audio-frequency) as the sound wave. An oscillator produces a continuous high-frequency (radio-frequency) current which has a fixed frequency of between 100 kilohertz and 1000 megahertz (1 kilohertz = 1000 cycles per second, 1 megahertz = 1 million cycles per second). This current produces the carrier wave. The audio-frequency current and the radio-frequency current are mixed in the transmitter so that the carrier wave is modulated by the audio-frequency current in exactly the same way as the sound fed into the microphone. If the signal has to travel a long way it is amplified before transmission.

A wave may be modulated in two different ways. Amplitude modulation (AM), is the method by which the amplitude or height of

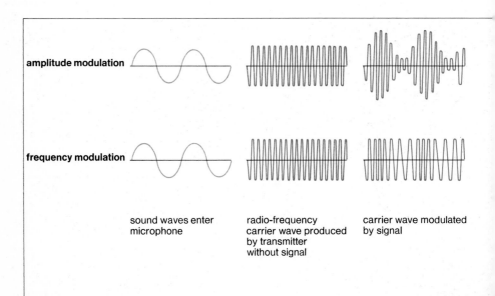

amplitude modulation

frequency modulation

sound waves enter microphone

radio-frequency carrier wave produced by transmitter without signal

carrier wave modulated by signal

the carrier is changed by the microphone current. When the modulated wave is picked up by a receiver the sound-frequency wave is separated by valves or transistors from the radio-frequency carrier. It is then amplified and the original sound reproduced by the loudspeaker. Different transmitters use carrier waves of different frequencies so the receiver must be tuned to pick up the carrier wave frequency of the required station.

Another method of modulation is frequency modulation (FM). The carrier wave has a much higher frequency (VHF = very high frequency) and a much shorter wavelength than for amplitude modulation. In this method amplitude is kept constant but the frequency is changed by the modulating microphone current.

Frequency modulation has two advantages over AM. It is less affected by interference and it reproduces music more faithfully. A VHF receiver is somewhat more complicated than one for amplitude modulation but the reception is very much better. However, VHF has a much shorter range than medium frequencies and is therefore used mainly by local radio stations.

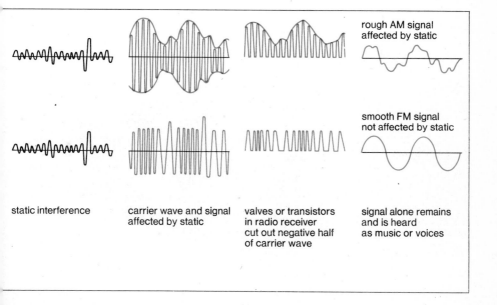

rough AM signal
affected by static

smooth FM signal
not affected by static

static interference

carrier wave and signal
affected by static

valves or transistors
in radio receiver
cut out negative half
of carrier wave

signal alone remains
and is heard
as music or voices

When a voltage is applied between two electrodes sealed into a tube from which most of the gas has been removed, a stream of electrons passes from the negative electrode (cathode) to the positive electrode (anode). Because the stream travels in straight lines, like light rays, they were called cathode rays. If the anode has a hole in it, the cathode rays will pass through it and strike the end of the glass tube producing a greenish glow. It was from the study of cathode rays that the Englishman, J. J. Thomson, discovered electrons in 1897. This discovery led to the development of the whole field of electronics.

The modern cathode ray tube uses a narrow beam of electrons produced by heating a filament similar to the one in a thermionic valve. The beam is focused by a system of cylindrical electrodes which act in the same way as a lens acts on a beam of light. The system of electrodes together with the filament is called an electron gun. The electrons travel at the enormous speed of about 80 000 km per second. The widened end of the tube opposite the electron gun is coated with chemicals (phosphors) which glow when struck by the electron beam and produce a green spot on the screen.

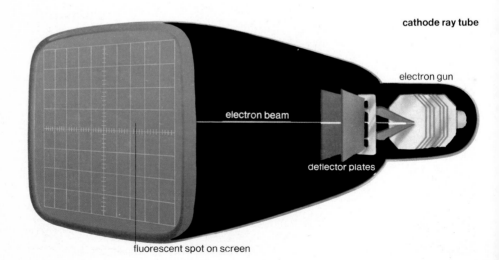

cathode ray tube

electron gun

electron beam

deflector plates

fluorescent spot on screen

As electrons are negatively charged they are attracted by a positive charge and repelled by a negative charge. Inside the tube are two sets of deflector plates. One set is vertical and the other horizontal. When voltages are applied to the deflector plates the electron beam, and thus the spot of light on the screen, can be moved up or down and from side to side. The amount of deflection is in proportion to the voltage applied to the plates. By applying voltages to both pairs of plates at the same time the spot can be moved to any point on the screen.

In a cathode ray oscilloscope a special circuit (time-base) is connected to the vertical plates which makes the spot move very rapidly across the screen from left to right. This movement can be repeated again and again in rapid succession. In this way a trace having the shape of a wave can be produced on the screen if an alternating current is applied to the horizontal plates. The cathode ray oscilloscope has many uses, especially in examining the output of electronic devices. It also forms the basis of the electrocardiogram for examining heartbeats.

Screen of a cathode ray oscilloscope

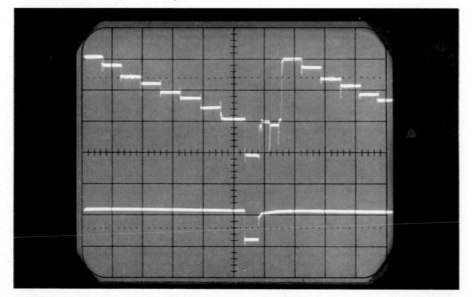

By the end of the 19th century pictures could be transmitted by telegraph. The method, called scanning, is similar to the way you are reading this page. Your eyes see a few words at a time and move from left to right along one line then jump back to the beginning of the next line and so on. To understand how a picture can be scanned think of it as being made up of hundreds of lines, each composed of small patches of varying shades of black and white. The reflections from these patches of dark and light are converted into a varying electric current, which can be transmitted by wire and reconstructed into the original picture by a receiver.

The first successful transmission of a picture by radio was made in London by John Logie Baird in 1926. He devised a mechanical method of scanning the picture in which a rotating disc of lenses produced successive images of each part of the picture. The picture was broken up into 240 lines and the final result was a flickering image. But to produce a sharper picture, more lines and faster scanning were needed.

This became possible when the mechanical scanner was replaced

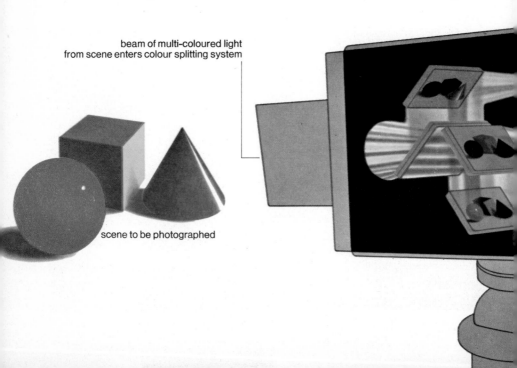

beam of multi-coloured light
from scene enters colour splitting system

scene to be photographed

by a more agile electronic scanner, the iconoscope, which was invented in 1938 by Vladimir Zworykin, a Russian working in America. The iconoscope made use of a form of photoelectric cell and the scanning was carried out by a moving electron beam similar to that used in the cathode ray tube. The iconoscope has been replaced in modern television cameras by the orthicon, which is a more sensitive device that operates on the same principle.

The camera focuses an image of the scene onto a mosaic made up of thousands of tiny photoelectric cells which become positively charged when light falls on them. The amount of charge is proportional to the brightness of the light. The cells are scanned by an electron beam which discharges each photoelectric cell in turn as it passes over it. The current carried by the beam depends on the brightness of each element and is used to modulate a radio-frequency carrier wave as in sound broadcasting, except that in this case the modulations correspond to the variations in light and shade of the picture. With a modern system using 625 lines, the scene is scanned 30 times in every second. Each complete scan is called a frame.

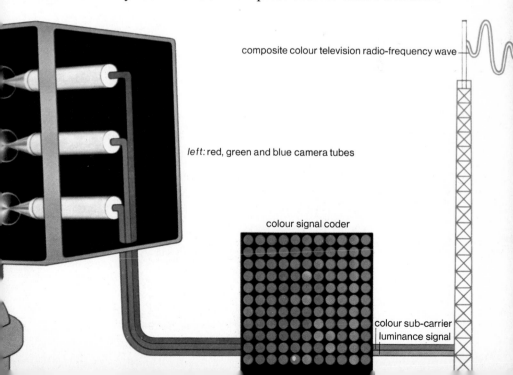

composite colour television radio-frequency wave

left: red, green and blue camera tubes

colour signal coder

colour sub-carrier
luminance signal

The picture in a television receiver is produced by a type of cathode ray tube with a very large screen at one end. An electron beam scans the screen line by line in exactly the same way as the original picture was scanned in the television camera. The beam is deflected magneti- cally by two pairs of deflecting coils, one for horizontal scanning and one for vertical deflection. These coils are connected to the horizontal and vertical time-base circuits. The modulated wave from the trans- mitter is received and fed to the grid of the electron gun in the cathode ray tube – this controls the number of electrons striking each spot of the phosphor and thus the brightness of the spot on the screen. As the spot scans across the screen line by line a complete picture is built up in each frame.

The scanning in the receiver must be exactly in step with the scanning in the camera, each frame starting and finishing at the correct instant. A series of synchronizing signals is transmitted to keep the camera and the receiver in step. The sound signal also has to be synchronized with the picture signals. Because a television receiver has to deal with much more information than a radio receiver

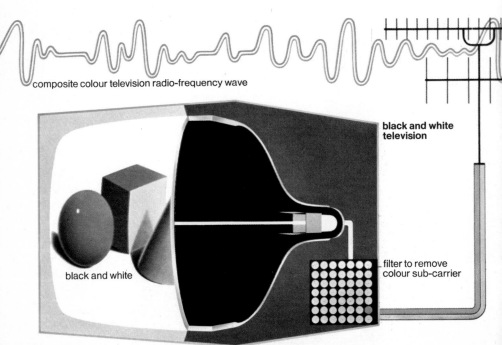

composite colour television radio-frequency wave

black and white
television

black and white

filter to remove
colour sub-carrier

it is a much more complicated instrument.

Colour television makes use of the fact that white light is a mixture of the colours of the spectrum and that all colours can be produced by mixing the three primary-coloured lights, red, green and blue, in the correct proportions. A colour television camera consists essentially of three cameras in one. Filters in the optical system of the camera split the light into its three colour components and each of the coloured images formed in this way is scanned by a separate electron beam. Three separate signals – red, green and blue – are then transmitted from the television station.

There are three electron guns in the tube of the colour receiver. The screen has three different phosphors, each one emitting either red, green or blue light. The whole screen contains about half a million dots arranged in groups of three's. The red, green and blue electron guns scan the red, green and blue dots respectively and thus build up a complete colour picture. Black-and-white receivers can use a signal which is equal to the sum of the three colour signals to receive a black-and-white picture of the colour transmission.

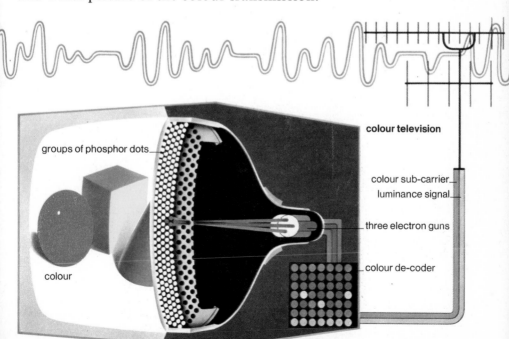

colour television

groups of phosphor dots

colour

colour sub-carrier
luminance signal

three electron guns

colour de-coder

The word radar comes from the initial letters of *ra*dio *d*etection *a*nd *r*anging. It is a system of detecting distant moving objects and measuring how fast they are travelling and in what direction. Radar works by sending out a beam of short radio waves which is reflected back to the transmitter by any object in its path. The distance of the object can be calculated from the time interval between sending and receiving the beam.

Radar was first used in England by Sir Edward Appleton in 1924 to detect electrically-charged layers in the upper atmosphere (ionosphere). It was later developed for military purposes by Sir Robert Watson-Watt. Radar differs from ordinary radio broadcasting in several ways. The transmission is not continuous but consists of a series of high-powered radio pulses, each lasting for a ten-millionth of a second. The pulses are sent out in a beam by placing a reflector behind the transmitting aerial. The beam can be sent in any single direction or the whole transmitting system can be rotated so that the beam sweeps a circle. The reflected signal comes back to the transmitter and is displayed on the screen of a cathode ray oscilloscope.

Most airports have a radar system to monitor the position of incoming aircraft

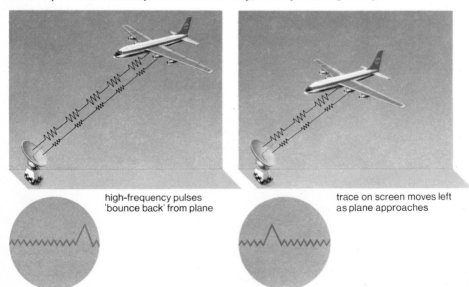

high-frequency pulses
'bounce back' from plane

trace on screen moves left
as plane approaches

Receiving equipment must be very sensitive because the fraction of the pulse reflected back is very small.

Medium-wave broadcasting uses wavelengths of several hundred metres. Early radar systems used wavelengths of about ten metres. Even these waves were too long to detect aircraft accurately, and transmitting and receiving aerials for this length of wave were too big to be carried in an aircraft. A new kind of valve, a magnetron, was invented in 1939 which was capable of producing waves only a few centimetres long. It enabled smaller and more sensitive radar sets to be developed for detecting enemy aircraft in the Second World War.

Ships now use radar because its beam works equally well in darkness and fog. Electrical disturbances in the atmosphere and rain clouds show up on a radar screen and this allows aircraft pilots to avoid them. Fast-moving rockets several thousand miles away can be detected by a missile early warning system. All modern airports and most aircraft are equipped with radar which enables pilots to land when weather conditions are bad.

Left: radar equipment at Hammersmith Bridge, London, scans reservoirs, Cromwell Road in the south-west and houses in the narrow streets of Fulham. Right: *map of the same area*

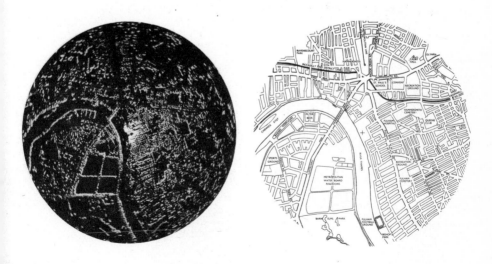

Radio waves travel outwards from a transmitting aerial in straight lines and do not follow the curvature of the Earth. Marconi was able to send a message across the Atlantic Ocean because there is an electrical layer in the upper atmosphere which reflects the radio waves back to Earth. In fact, several such electrical layers have since been discovered. The Heaviside-Kennelly layer, or E-region, is about 90 to 150 km above the Earth's surface and the Appleton layer or F-region lies between 150 and 400 km. Both these layers consist of ionized gases and form part of the ionosphere. Radio waves, which travel by reflection from the ionosphere, are called sky waves.

Long radio waves are reflected by the E-region and short waves by the F-region. However, the very short waves of VHF (very high frequency) and UHF (ultra high frequency) transmissions are not reflected but pass through the ionosphere into space. Television programmes which are broadcast on these high frequencies, therefore, cannot be reflected from the ionosphere. Another disadvantage of the ionosphere is that heights and densities of the layers constantly change, particularly during magnetic storms and periods of sunspot

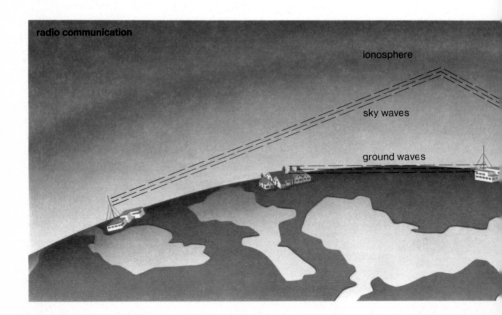

radio communication

ionosphere

sky waves

ground waves

activity. This makes long-distance transmissions fade and can cause a complete radio blackout.

An artificial reflector does not have these disadvantages. Early communication satellites such as the echo balloons were simply large balloons with metallic surfaces to reflect the radio waves. But the signals were scattered by the balloon in all directions and much of the power was lost. Telstar, an active satellite launched in July 1962, contains more than 2500 transistors. Signals beamed to it from one side of the Atlantic Ocean are re-transmitted to the receiving aerial on the other side.

Early Bird satellites, also active, are put into stationary orbits about 35 000 km above the Earth. One orbit takes exactly 24 hours so each orbit keeps pace with the Earth's rotation. In this way the satellite remains vertical above the same spot on the Earth and is always available for re-directing television programmes from one area of the Earth to another. Communication satellites are powered by solar cells which charge storage batteries when the satellite is not transmitting.

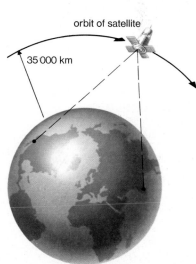

satellite communication

orbit of satellite

35 000 km

Some of the early Greek philosophers, including Democritus and Epicurus, thought that matter might be made up of very small particles and that the characteristics of different substances depend on how their particles are arranged. The word 'atom' in Greek means indivisible and the hypothesis was that everything is made up of atoms, but that the atoms themselves cannot be split into anything simpler. The Greeks were not scientists and did not carry out experiments to try to prove their ideas.

It was not until experiments were made with gases in the 17th and 18th centuries that the atomic theory could be given a scientific basis. These experiments and the discovery of the simple laws of chemical combination helped the English chemist, John Dalton, to formulate his Atomic Theory of Matter in the early 19th century. During this period scientists pictured an atom as a very tiny, hard, round particle like a minute billiard ball.

This simple view of matter was considerably changed in 1897. J. J. Thomson, the British physicist, discovered that the cathode rays,

Left: *Lord Rutherford pioneered the development of sub-atomic physics.* Right: *John Dalton's Atomic Theory of Matter elevated chemistry into a science*

produced when a high-voltage discharge is passed through a low-pressure gas, are streams of negatively charged particles (electrons). Thomson found that no matter what gas was in the tube the electrons always behaved in the same manner. He concluded that the electron forms a part of every kind of atom. Because atoms are electrically neutral and electrons are negatively charged, it followed that atoms must also contain some positive charge. Thomson thought that the negative electrons were embedded in a larger positive charge.

In the early 1900s Lord Rutherford, the physicist from New Zealand, disproved this theory while working in England. By bombarding gold atoms with alpha-particles he showed that from the way in which they are deflected, the positive charge of the atom is concentrated in a central core – the nucleus. The electrons are outside this core and circle around it, like the planets around the sun. He was able to show that the central nucleus contained the same number of positively charged particles (protons) as there were orbiting electrons. As the charge of the proton is equal, but opposite, to that of the electron, the complete atom is neutral.

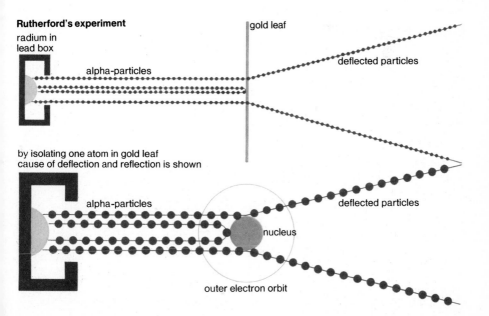

Rutherford's experiment

radium in
lead box

gold leaf

alpha-particles

deflected particles

by isolating one atom in gold leaf
cause of deflection and reflection is shown

alpha-particles

deflected particles

nucleus

outer electron orbit

Rutherford's suggestion that the positive charge of the atom is concentrated in a small central nucleus was a revolutionary one. With a Danish physicist, Niels Bohr, he worked out a model of the atom which in many respects is still accepted today.

Their model relies on three different particles: (1) the proton, with unit positive charge and a mass about 1840 times greater than the electron, (2) the neutron, which is uncharged and has a mass practically equal to the proton mass and (3) the electron, with a unit negative charge. The nuclei of atoms are made up of protons and neutrons except for hydrogen which consists of one proton. The planetary electrons are grouped together in various orbits; each group of electrons is called an electron shell.

The complete atom is electrically neutral so that the number of electrons orbiting the nucleus is equal to the number of protons in the nucleus; this number is called the atomic or proton number. The mass of the atom is made up of the neutrons and the protons in the nucleus (the two particles are often called nucleons). The number of protons and neutrons added together gives the mass number or

Left: *Niels Bohr outlined a new atomic theory.* Right: *James Chadwick discovered the neutron*

nucleon number. A hydrogen atom has a nucleus of one proton and has one electron outside it. The atomic number is one and the mass number is one. Carbon has six protons and six neutrons in the nucleus and six orbiting electrons. The atomic number is six and the mass number is 12. The first electron shell contains two electrons and then it is 'full'. The second shell can hold eight and the third 18. A more complex picture of the atom would also include an analysis of sub-shells.

The chemical properties of the atom depend on the number of planetary electrons, particularly those electrons in the outer shell. It is these outer electrons that take part in chemical reactions. Although all the atoms of an element have the same number of protons in the nucleus, they do not all have the same number of neutrons. Atoms with different numbers of neutrons are called isotopes. For example, deuterium and tritium are isotopes of hydrogen. Hydrogen has one proton in its nucleus and one orbiting electron, but deuterium has a nucleus consisting of one proton and one neutron while the tritium nucleus has two neutrons in addition to a proton.

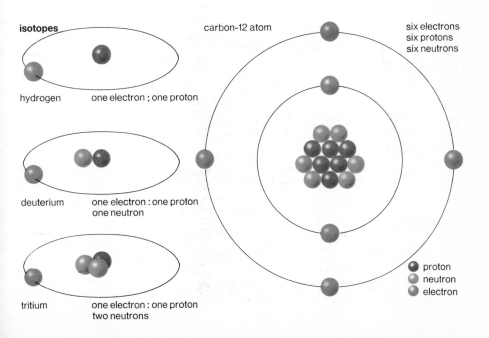

isotopes

carbon-12 atom

six electrons
six protons
six neutrons

hydrogen   one electron : one proton

deuterium   one electron : one proton
                 one neutron

tritium   one electron : one proton
                 two neutrons

proton
neutron
electron

X-rays are electromagnetic radiations consisting of waves with a shorter wavelength than ultraviolet radiation. They are produced when high-speed electrons hit a metal target. They were accidentally discovered in 1895 by the German physicist, Wilhelm von Röntgen, who noticed that certain chemicals glowed when placed near a cathode ray tube. He concluded that the tube must be giving out some form of invisible radiation. He also found that this radiation could pass easily through wood and paper.

X-rays are useful because they can travel through some substances but are absorbed by others. They pass through flesh, for example, but are stopped by bone, and as they have the same effect on photographic plates as light, they can photograph bones. This is done by placing a photographic plate beneath the part of the body to be examined and exposing it to a source of X-rays. When the plate is developed, dark areas corresponding to the bones will appear, which clearly show any damage. Diseased tissue, such as a patch of tuberculosis on the lung, can also be detected by X-rays because it is more dense than healthy tissue. In industry, X-rays detect flaws in metal

The X-ray photograph immediately reveals if there is any bone damage in these hands

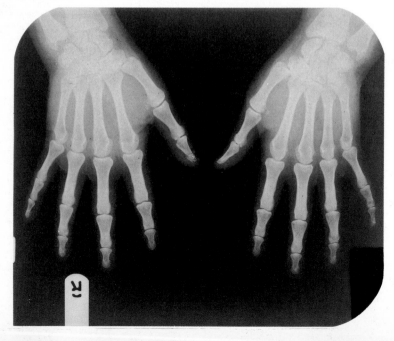

castings and welded joints.

Electrons are produced in a modern X-ray tube by heating a cathode filament in a vacuum. The electrons are accelerated by a high voltage up to 100 000V so that they strike a tungsten anode. When the electron beam strikes the tungsten atoms, the inner electrons in the tungsten atoms are disorganized and may jump from one electron shell to another, moving away from the nucleus. When they fall back to the original shell, X-rays are given out.

In 1914, by measuring the wavelengths of the X-rays emitted by different kinds of atoms, the English physicist, Henry Moseley, evolved the idea of atomic numbers. Thus, he made sense of Mendeleyev's early form of the periodic table, and was able to predict the existence of elements that had not yet been discovered.

X-rays can damage living cells by disturbing the arrangement of electrons in the atoms and thus producing chemical changes. Exposure to X-rays, therefore, must be for short periods only. Cancer cells are killed more easily by X-rays than healthy cells, and therefore X-rays are often a help in the control of cancer.

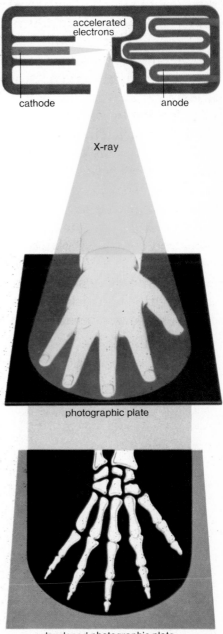

accelerated electrons

cathode

anode

X-ray

photographic plate

developed photographic plate

A radioactive element is one in which a proportion of the nuclei of the atoms spontaneously breaks up, forming new atoms and emitting radiations. The rate at which the atoms disintegrate or decay varies considerably from isotope to isotope, but the rate remains the same for a particular isotope and no alteration of physical conditions, such as temperature, can change it.

Radioactivity was discovered in 1896 by the French physicist, Antoine Becquerel. He noticed that a photographic plate kept in a drawer mysteriously showed an image of a key when developed. Although the key was lying on the plate in the drawer, the plate itself was kept in black paper to exclude the light. After careful experiments Becquerel concluded that a sample of uranium ore in the drawer was giving off some form of invisible radiation which could pass through paper and even metal plates. The Frenchman, Pierre Curie, and his Polish wife, Marie, later discovered the radioactive element, radium.

There are three kinds of radiations given off by radioactive substances. Alpha-particles are the nuclei of helium atoms. They are positively charged because they consist of two protons (positive) and

Natural radioactivity used to take a photograph. Pitchblende, which contains radium, is placed over a metal key resting on a photographic plate wrapped in black paper. A picture of the key appears on the developed plate.

two neutrons (uncharged). Beta-particles are electrons. They do not exist in the nucleus but are emitted by it when a neutron in the nucleus changes into a proton and an electron. Gamma-rays are very short wavelength electromagnetic waves which are even shorter than X-rays. They are the most penetrating of the radiations and can pass through several centimetres of lead.

A measure of the activity of a radioactive element is its half-life period. This is the time it takes for half of the atoms present to decay. The half-life period of the most stable isotope of radium is 1620 years, and for one of the isotopes of plutonium it is only 0·2 seconds. We can never say which atoms in a particular sample of radioactive material will disintegrate, but when the half-life has been measured we know that half the number of atoms originally present will have changed. In a second half-life period the original number of atoms will have been reduced to a quarter and so on. Thus the break-up of a nucleus is a statistical process – like the probability of death. We know fairly accurately the total number of people who will die in this country next year, but we cannot say who they will be.

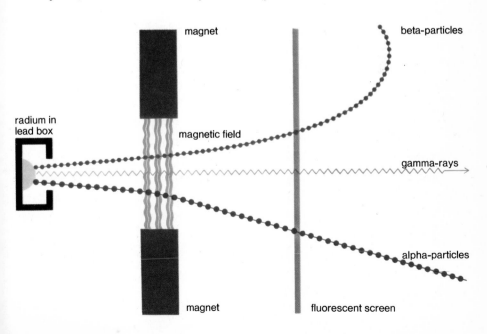

When radioactive radiation passes through a gas it detaches some of the outer electrons from the atoms of the gas. This is called ionization, and as the gas then contains some free electrons it is able to conduct electricity. This current in a gas is one way to detect the presence of radioactive material. A Geiger counter consists of a glass tube containing an inert gas, such as argon, at low pressure. Within the tube are two electrodes (the anode is usually a wire running along the axis of a cylindrical cathode) between which a high voltage is maintained. When alpha-, beta-, or gamma-radiation enters the tube the gas is momentarily ionized and the pulse of current created can produce a click in a loudspeaker or work an electronic counter.

Radioactive substances are dangerous. Alpha- and beta-particles cannot travel far in living tissue and mainly cause radiation burns on the skin. Gamma-rays, however, are much more dangerous because of their high energy and penetrating power. They can damage the molecules of genetic material (DNA) stored in the reproductive organs causing mutations which will be harmful to future generations. Substances emitting gamma-radiation are therefore kept behind thick concrete and lead shields.

The constant rate at which radioactive atoms disintegrate provides a very useful way of measuring time. For example, the age of radioactive rocks can be estimated by measuring their present composition. Radioactive uranium-238 decays through a whole series of radioactive products to ordinary lead. The half-life of uranium-238 is $4 \cdot 5 \times 10^9$ years. Assuming an ore originally contained pure uranium and that its present composition is half uranium and half lead, we could assume that the sample was $4 \cdot 5 \times 10^9$ years old. Organic material such as wood can be dated by measuring its content of carbon-14, a rare radioactive isotope of carbon. This isotope occurs in the atmosphere in tiny amounts and therefore while a tree is alive it takes in C-14 during photosynthesis. When it is cut down it ceases to acquire this isotope and the amount of C-14 stored in it decreases due to radioactive decay. Therefore the amount of C-14 in a wooden specimen indicates when the tree was cut down. This method will date wood or bone up to an age of about fifty thousand years.

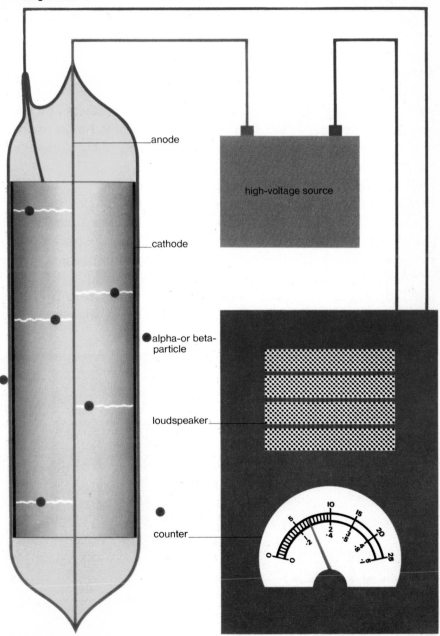

Geiger tube

anode

high-voltage source

cathode

alpha-or beta-particle

loudspeaker

counter

The alchemists of the Middle Ages searched unsuccessfully for a way to turn common metals into gold. The chemists of the 19th century, with their ideas of the fixed nature of atoms, thought that the transmutation of one element into another was impossible. But the discovery of radioactivity altered this view, because radioactive atoms are continuously emitting fragments of their nuclei and changing into different atoms with different atomic numbers and atomic masses. These radioactive changes occur naturally, but at the beginning of this century scientists began experiments to try to change one element into another artificially.

Lord Rutherford, working in the Cavendish Laboratory at Cambridge University in 1919, tried bombarding nitrogen atoms with alpha-particles. He hoped that a few alpha-particles might hit the nuclei of nitrogen atoms, either chipping bits off the nuclei or combining with them. If either of these processes took place new atoms would be formed. Using very simple apparatus Rutherford found that one result of the collisions was the production of protons (hydrogen nuclei). In 1925 P.M.S. Blackett, an English physicist,

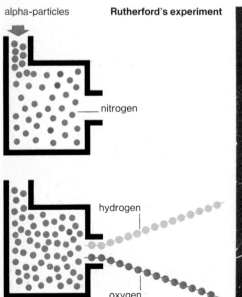

alpha-particles   **Rutherford's experiment**

nitrogen

hydrogen

oxygen

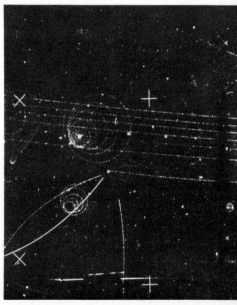

took more than 20 000 cloud chamber photographs of the atom tracks of this kind of collision process. He showed that some of the alpha-particles had, in fact, combined with the nitrogen nuclei which caused a proton to be emitted, and formed the nucleus of an isotope of oxygen in the process. This was the first time that one element had been changed into another. Since then, many nuclear transformations have been carried out, including the changing of mercury into gold. Nuclear transformations have also enabled scientists to produce 13 new elements with heavier atoms than uranium.

The cloud chamber is an instrument that makes the tracks of atomic particles visible. It was invented by C.T.R. Wilson, a Scottish physicist, in 1911. The tracks are produced in much the same way as the vapour trails of high-flying aircraft. As the atomic particles pass through the chamber, they ionize the air which has been saturated with water vapour. The air is suddenly expanded which causes cooling and water drops to form on the ions. The paths of the particles then become visible as a row of droplets. The particles can be identified by the lengths and intensities of their tracks.

Cloud chamber photographs provide information on particle reactions

The first attempts to smash the atom used alpha-particles emitted by radium as the bombarding particles. But the rate at which the alpha-particles are fired cannot be controlled and, because they are positively charged, they are repelled by the positively charged nucleus. Therefore the chances of obtaining a direct hit are small unless they are accelerated to a very high velocity. Beta-particles (electrons) from radioactive sources make poor ammunition because they have a small mass and to be effective they, too, must be accelerated.

Particle accelerators are machines for producing fast-moving (and therefore high-energy) protons and electrons for breaking up nuclei and causing nuclear reactions. In an accelerator a stream of protons is accelerated between positively and negatively charged electrodes, and thus is repelled by the positive electrode and attracted by the negative electrode. The greater the voltage between the electrodes the greater the force and the greater the acceleration. The energy of the moving particle is measured in electron-volts (eV); this is the energy of a single, charged particle (electron or proton) accelerated by a potential difference of one volt. Larger units are the MeV (mega or

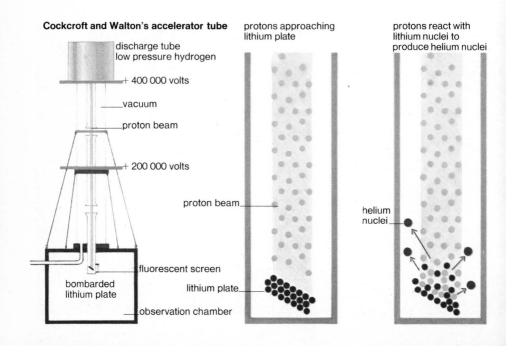

Cockcroft and Walton's accelerator tube

discharge tube
low pressure hydrogen

+ 400 000 volts

vacuum

proton beam

+ 200 000 volts

bombarded
lithium plate

fluorescent screen

lithium plate

observation chamber

protons approaching
lithium plate

proton beam

protons react with
lithium nuclei to
produce helium nuclei

helium
nuclei

million electron-volts) and GeV (giga or thousand million electron-volts): in the United States this is written BeV (billion electron-volts).

In 1932 John D. Cockcroft and Ernest Walton working at Cambridge University accelerated protons through a distance of about two metres by about 400 000 V and bombarded a lithium target. The protons reacted with the lithium nuclei to form helium nuclei ($^1_1H + ^7_3Li = 2\,^4_2He$). This was the first linear accelerator, so named because the accelerated particles travel in a straight line between a series of negative and positive electrodes.

Another type of particle accelerator is the cyclotron. Particles are made to travel by means of a powerful magnetic field in a spiral path between two D-shaped electrodes. An alternating voltage speeds up the particles by giving them a push in each of the Ds. The proton synchrotron at the Conseil Européen pour Recherches Nucléaires (CERN) in Geneva uses two colliding beams of particles to increase the energy. Particle accelerators are used in experiments to investigate the structure of nuclei and to produce radioactive isotopes. They have now become extremely expensive to operate.

At CERN, near Geneva, research is carried out on the fundamental structure of matter

Protons and alpha-particles can be accelerated to high speeds in particle accelerators but being positively charged they are repelled by positively charged nuclei. This reduces the chance of making direct hits. Neutrons with no charge are not repelled by nuclei. The chances of hitting a nucleus with a neutron are much greater.

In 1934 the Italian physicist, Enrico Fermi, used a stream of neutrons to bombard the nuclei of a radioactive isotope of uranium, uranium-235. He found that when a neutron struck a U-235 nucleus, the nucleus split into two parts producing two lighter elements, each having about half the mass number of U-235. In 1938 Lise Meitner, an Austrian physicist, and Otto Hahn, a German physical chemist, found that when the uranium nucleus splits into two parts there is a great release of energy and two or sometimes three additional neutrons are produced at the same time. These new neutrons can cause further uranium nuclei to split, resulting in a continuous chain reaction. The splitting of heavy atoms in this way is nuclear fission.

The release of energy occurs because the nuclei formed by the fission have less mass than the original uranium nucleus. Some of the

fission a fast-moving neutron collides with the nucleus of an atom of uranium-235 the neutron sets up oscillations in the nucleus and distorts it the stresses within the nucleus break it in two releasing energy in the form of an explosion two lighter nuclei and three free neutrons result from the explosion

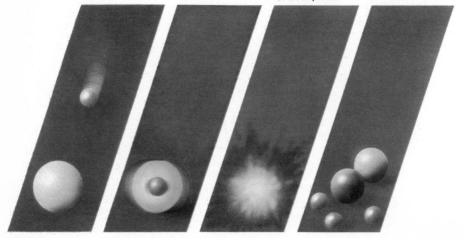

mass is converted into energy according to Albert Einstein's equation $E = mc^2$. The energy is released mainly as kinetic energy of the nuclei formed by the fission, which fly apart at great speed. Uncontrolled, this energy provides the destructive power of a nuclear bomb; under control it provides the source of power in a nuclear reactor.

Fission is the splitting of a large nucleus into smaller parts. Fusion is the joining together of two light nuclei to form one heavier nucleus. In fusion reactions there is also a loss of mass and an equivalent release of energy. Reactions of this type account for the production of energy in the sun and other stars. It is also the basis of the hydrogen bomb in which there is an enormous release of energy when deuterium (heavy hydrogen) nuclei combine to form helium. The deuterium nuclei will not combine unless they are moving rapidly enough to overcome the repulsive forces between them. To achieve this they have to be raised to an extremely high temperature, about 100 million °C. This process is thermonuclear fusion. Scientists are still striving to produce these tremendous temperatures under controlled conditions so that fusion energy can be released for peaceful purposes.

fusion two hydrogen isotopes deuterium and tritium collide the protons and neutrons of the two isotopes combine energy is released in re-arrangement causing an explosion the resulting components are a stable helium atom and a neutron

Nuclear physics
131   reactors

The fission of uranium-235 nuclei when bombarded with neutrons releases energy and two or three more neutrons. Some of these fission neutrons will be travelling so rapidly that they escape from the mass of uranium altogether. Others will hit more uranium nuclei to produce fission, energy and more neutrons, so that the process builds up into a chain reaction. This is the basis of the nuclear reactor in which the energy released is used as a source of heat. The reactor is like a furnace using uranium as the fuel. But it is a very powerful fuel – one kilogram of uranium-235 can produce as much heat as 2000 tons of coal.

Natural uranium contains two principal isotopes, uranium-238 (99·28%) and uranium-235 (0·71%). These isotopes behave differently when struck by neutrons. If natural uranium is used as the reactor fuel many of the fast neutrons created in the fission process will be captured by the U-238 nuclei to form Plutonium-239. These captured neutrons cannot create new fissions so the chain reaction stops. This is overcome by slowing down the neutrons, because they can then split the U-238 nuclei. These slower neutrons are also best for

thermal reactor

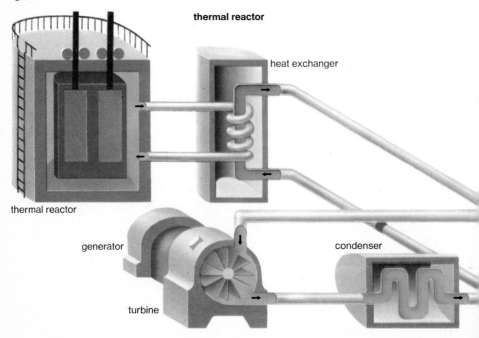

thermal reactor

heat exchanger

generator

condenser

turbine

splitting the U-235 nuclei. Neutrons are slowed down by mixing a moderator with the fuel. A moderator is a light substance, such as graphite or heavy water, with which the neutrons can make frequent collisions and so lose some of their energy. Reactors which use a moderator are thermal reactors.

Fast reactors do not use a moderator. Instead they use a fuel enriched with extra quantities of U-235 or Pu-239. The fuel is enriched just enough for the fast neutrons to maintain the chain reaction. To control the rate of reaction in a reactor, rods of boron or cadmium are raised or lowered into the fuel to absorb the neutrons.

The heat produced in the reactor is collected by a circulating coolant and passed to a heat exchanger in which water is converted into steam to drive a turbine. The turbine drives a generator to produce electricity or supply the power for a ship.

The American submarine *Nautilus*, launched in 1955, was the first ship to use nuclear power. The great advantage of nuclear power for submarines is that the reactors need no air so the vessel can remain submerged for long periods.

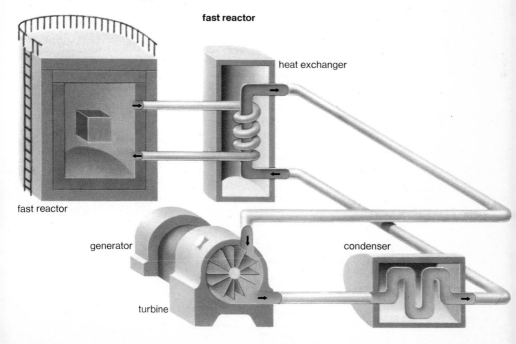

fast reactor

heat exchanger

fast reactor

generator

condenser

turbine

A few radioactive elements are found in the Earth's crust. Artificial radioisotopes (radioactive isotopes) can be produced by bombarding stable elements with protons or alpha-particles in a particle accelerator or with neutrons in a nuclear reactor. The first radioisotope was phosphorus-30, produced by Irene Joliot-Curie, daughter of Pierre and Marie Curie, and her husband Frederic in 1934. Since then at least one radioisotope of most of the elements has been produced.

As they decay, radioisotopes emit alpha-, beta- and gamma-radiation which can be detected by a Geiger counter. Even minute traces of the radioisotopes are shown on this very sensitive instrument. Radioisotopes provide a very useful method of tracing the path of an element through a plant or an animal. For example, to find out what happens to ordinary phosphorus when it is absorbed by plants, radioactive phosphorus-32 is added to the phosphate fertilizer fed to the plants. Chemically the P-32 atoms behave in exactly the same way as the non-radioactive P-31 atoms in the fertilizer – they are absorbed by the roots and travel through the plant. The P-32 acts as a radioactive tracer and the presence of phosphorus in the leaves can be

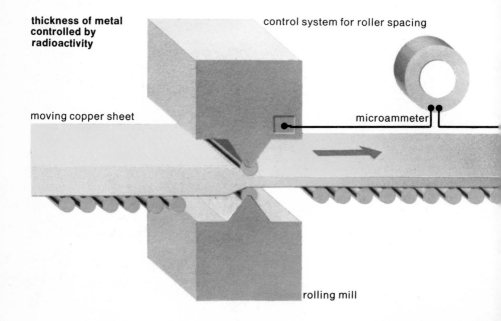

thickness of metal controlled by radioactivity

control system for roller spacing

moving copper sheet

microammeter

rolling mill

detected either with a Geiger counter or by placing a photographic film on the leaf. Photographic film is sensitive to radioactivity and the picture produced in this way is an autoradiograph. The monitoring of radioactively-labelled atoms can detect a leak in a pipe, check the movements of sandbanks in an estuary or measure the amount of wear in an engine.

Radioisotopes have other uses. If a radioisotope is placed on one side of a metal plate, the amount of radiation passing through it will depend on the thickness of the metal. So if a Geiger counter is placed on the other side of the metal its thickness can be checked to see if it is constant. The gamma-radiation from radioisotopes, such as cobalt-60, is used in radiotherapy for the treatment of tumours in the body which cannot be removed by surgery.

The radiation from radioisotopes is dangerous unless strict safety precautions are observed. The isotopes are handled with long tongs or remotely controlled mechanical hands and stored in thick lead containers. If they are sent by air they are carried in special compartments in the wing tips of the aircraft.

A shielded cell is necessary when handling compounds containing radioactive ingredients

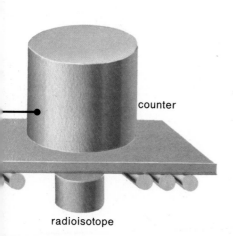

counter

radioisotope

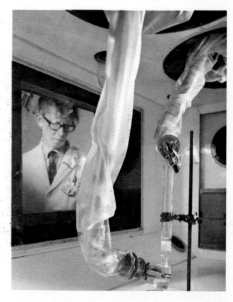

Chemistry
133 alchemists to atoms

Chemistry is the branch of science concerned with the composition of materials and how one substance can be changed into another. In recent times chemists have worked on the problems of making completely new materials with new properties. Plastics, for example, do not occur naturally – they are man-made materials. So, too, are stainless steel, nylon, petrol and aspirin.

The early Egyptians and Assyrians applied themselves mainly to such practical problems as the smelting of metals from their ores, the preparation of dyes for cloth, making glass and developing coloured glazes for pottery. Gold was the first metal they found, then came copper and silver, followed by tin, iron, lead and mercury. The early chemists were called alchemists. To explain chemical changes they based their theories on the imaginary properties of what they called the four 'elements', Earth, air, fire and water. They hoped to produce a 'philosopher's stone' which would turn base metals into gold, and the 'elixir of life' which would make people live forever. Although they failed to produce any of them, the alchemists built up a practical knowledge of chemical processes and a method of performing

An Egyptian tomb painting of 1425 B.C. shows the method used for casting a bronze door

experiments that was later to form the basis of modern chemistry.

Chemistry as a true science began in the 18th century with the study of gases and burning. The properties of gases such as hydrogen and oxygen were carefully studied and the use of sensitive balances helped chemists to measure accurately the changes in weight which accompany chemical changes.

By the beginning of the 19th century two important laws of chemistry had been established: (1) the law of conservation of matter which says that matter can neither be created nor destroyed, and (2) the law of constant composition which summarizes the results of many experiments and states that a chemical compound always contains the same constituents in the same proportions by weight. For example, water always consists of $\frac{1}{9}$ hydrogen and $\frac{8}{9}$ oxygen by weight no matter where it comes from. This discovery provided the first experimental evidence for the theory that all matter consists of atoms. This is an example of the way in which modern science relies on experiments – no theory is accepted until precise measurements show that it gives accurate predictions.

A 17th-century Dutch painting of an alchemist trying to change base metals into gold

element
sulphur

element
iron filings

iron and sulphur are elements which when heated combine to make the compound iron sulphide. When mixed without heat they remain easily separable.

mixture

compound

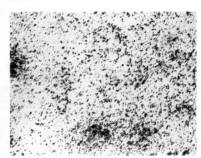

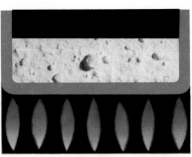

sulphur and iron filings

iron sulphide

An element is a substance that cannot be split into a simpler substance. When an electric current is passed through water two gases, hydrogen and oxygen, are produced. But no amount of chemical treatment can split hydrogen into anything simpler. This is also true of oxygen and the 92 other substances (elements) that make up all the different materials found in the Earth, planets and stars.

Water is a compound because it consists of two elements combined together chemically. It is only one of several million different compounds that exist as a result of two or more elements combining in different proportions and different ways. Compounds are only formed when elements react chemically with each other. Hydrogen and oxygen gases can be mixed together and they remain separate substances. But if the mixture is ignited an explosive chemical reaction takes place and the compound, water, is formed. A chemical reaction between elements produces compounds with different properties to the reacting elements. For example, common salt is a compound of the elements sodium and chlorine. Although chlorine is a poisonous green gas and sodium is a soft yellow metal which reacts very violently with water, the compound of these two elements is salt, the harmless white crystalline substance (sodium chloride).

However, when substances are mixed together without combining chemically they keep their own properties. A mixture of salt and sand is yellowish-white in colour; it tastes both salty and gritty and the salt can be dissolved out in water. Air is the most common mixture consisting of nitrogen and oxygen together with a small percentage of carbon dioxide and traces of such rare gases as helium and neon.

Elements consist of atoms, which are the smallest particles of an element that can exist and still show the characteristic properties of the element. The Atomic Theory of Matter to explain the formation of compounds was first put forward by the English scientist, John Dalton, in the early 19th century. The atoms of the 92 naturally occurring elements are all different from each other. Chemists have now made another 13 elements by bombarding existing elements with atomic particles, but these new atoms are very unstable and rapidly split up again into stable elements.

Dalton's Atomic Theory provided a simple explanation of the formation of compounds and the law of constant composition. Elements exist as atoms, and when atoms combine together they form compounds. The smallest part of a compound that can exist by itself is a molecule. For example, a molecule of water consists of two atoms of hydrogen and one atom of oxygen. A molecule of sugar (sucrose or cane sugar) consists of 12 carbon atoms, 22 hydrogen atoms and 11 oxygen atoms. The very large molecules of proteins found in living cells may contain more than a hundred thousand atoms.

Dalton represented the atoms of elements by pictorial symbols, but letter symbols are now in use. For example, the symbol H stands for one atom of hydrogen and O stands for one atom of oxygen. The atomic composition of a molecule of a compound can be shown by writing the symbols of the elements of which it is made with a small figure to show the number of atoms present. This gives the chemical formula of the compound. The formula of water is H_2O; for sucrose it is $C_{12}H_{22}O_{11}$. Because atoms of gases, such as hydrogen and oxygen, usually exist in pairs, the formulae for hydrogen and oxygen

structure of sucrose

(as gases) are H_2 and O_2 respectively.

The formation of a chemical compound can be represented by a chemical equation. The equation for the formation of water is $2H_2 + O_2 = 2H_2O$. Two molecules of hydrogen combine with one molecule of oxygen to form two molecules of water. Since atoms cannot disappear during a reaction (law of conservation of matter) there must be the same number of atoms of each element on both sides of the equation to make the equation balance. Chemical equations can be used to work out the weights of the reacting substances. Equations may be very complicated. For example, to make soap by boiling a fat with caustic soda is a simple process that has been known for about two thousand years. The equation for the chemical reaction is $C_{57}H_{110}O_6 + 3NaOH = C_3H_8O_3 + 3NaC_{18}H_{35}O_2$. When you check the equation you will find that on each side there are 57 carbon (C) atoms, 113 hydrogen (H) atoms, 9 oxygen (O) atoms and 3 sodium (Na) atoms. Many chemical equations can be written down but not all of them represent actual reactions – a true reaction can only be established by experiment.

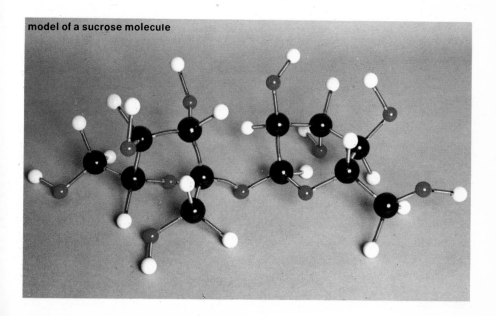

model of a sucrose molecule

According to Dalton's Atomic Theory, all the atoms of each element have identical properties and identical weights, while atoms of different elements have different weights. Dalton thought it would be useful to compare the weights of different atoms. Although his experimental results were not very precise, and his ideas about the numbers of atoms combining together to form a molecule of a compound were sometimes wrong, his general theory was correct. Using modern information we know that water is $\frac{1}{9}$ hydrogen and $\frac{8}{9}$ oxygen by weight. Therefore, one part of hydrogen combines with eight parts of oxygen; or two parts of hydrogen combine with 16 parts of oxygen. Since two atoms of hydrogen combine with one atom of oxygen (formula H_2O) it follows that one atom of oxygen is 16 times heavier than one of hydrogen. Using this principle, the atomic weight of an element was originally expressed as the weight of its atom compared with an atom of hydrogen, the lightest of the elements. Atomic weights are thus comparative weights. Modern atomic weights are compared with the weight of a carbon atom taken as 12, which makes the atomic weight of hydrogen 1·007 97, but the

When John Dalton lectured in 1806–7 he used this diagram of the elements and their weights

	W.ᵗ			W.ᵗ
Hydrogen.	1		Strontian	46
Azote	5		Barytes	68
Carbon	54		Iron	50
Oxygen	7		Zinc	56
Phosphorus	9		Copper	56

principle remains the same. Although the weights of individual atoms are now known, for example, 1700 million million million hydrogen atoms weigh only one milligram, the relative weights still remain the more useful information.

At the beginning of the 19th century about twenty-five elements were known and by 1850 the number was nearly sixty. A logical way of arranging the elements in order of increasing atomic weights was suggested by the Russian chemist, Dmitri Mendeleyev, in 1869. A pattern emerged – the periodic table – because the elements fell into groups with similar properties. At first there were gaps and a few elements did not appear in quite the right places, but today all the elements have been discovered and all the apparent mistakes explained.

Contrary to Dalton's original theory it has been found that all atoms of the same element do not have exactly the same weight although they have the same chemical properties. This is because all the atoms of the same element contain the same number of protons but some atoms contain different numbers of neutrons. Atoms of the same element having different weights are isotopes.

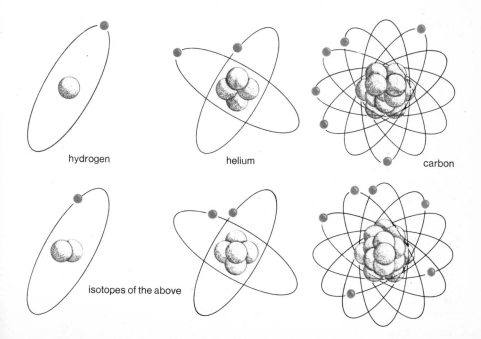

hydrogen   helium   carbon

isotopes of the above

Chemistry
137   a logical table of the elements

Mendeleyev arranged the elements into eight vertical columns, in order of increasing atomic weight, so that elements with similar properties fell into the same column. For example, the first column contained the alkali metals lithium, sodium and potassium, while the eighth column contained the inert gases helium, neon and argon. He found that basing the order on atomic weights created certain discrepancies. Argon, for example, to be in the same column as helium must come before potassium, but the atomic weight of argon is approximately forty, while potassium is about thirty-nine.

Differences of this kind were not resolved until 1914 when the English physicist, Henry Moseley, examined the X-rays emitted by the atoms of the elements and found a regular decrease in wavelength as he went through the periodic table in the correct (not the atomic weight) order. It was decided to number the elements, irrespective of atomic weight, starting with hydrogen at number one and finishing with uranium at 92, with gaps left for undiscovered elements. It has now been found that these atomic numbers are of great structural importance. The atomic number of an element is equal to the number of protons in the atomic nucleus (sometimes called proton number) and it is also equal to the number of electrons circulating around the nucleus. All the atoms of an element have the same proton number, but different isotopes have different mass numbers (the number of protons plus the number of neutrons in the nucleus). So the atomic number is equal to the mass number minus the neutron number.

In the case of argon, atomic number 18, there are three isotopes with mass numbers 40 (22 neutrons), 38 (20 neutrons) and 36 (18 neutrons). The average of these mass numbers taken in the proportions in which the isotopes occur in nature, gives an atomic weight of 39·95. Potassium, on the other hand, has isotopes with mass numbers 39, 40 and 41 (atomic number 19). Their average is 39·10. The atomic numbers are in ascending order, but the atomic weights are not. All the discrepancies have now been explained. To fit in the complicated transition elements, with their incomplete inner electron shells, it is necessary to subdivide the columns into A and B subgroups.

Mendeleyev made the first arrangement of the elements according to their atomic weights

alkali and other alkaline earth metals		first transition metals					second transition m
group 1a	group 2a	group 3b	group 4b	group 5b	group 6b	group 7b	group 8
1 H hydrogen 1.00797							
3 Li lithium 6.939	4 Be beryllium 9.0122						
11 Na sodium 22.9898	12 Mg magnesium 24.305						
19 K potassium 39.102	20 Ca calcium 40.08	21 Sc scandium 44.956	22 Ti titanium 47.90	23 V vanadium 50.942	24 Cr chromium 51.996	25 Mn manganese 54.938	26 Fe iron 55.847	27 Co cobalt 58.933
37 Rb rubidium 85.47	38 Sr strontium 87.62	39 Y yttrium 88.905	40 Zr zirconium 91.22	41 Nb niobium 92.906	42 Mo moly– bdenum 95.94	43 Tc technetium 97	44 Ru ruthenium 101.07	45 Rh rhodiu 102.90
55 Cs caesium 132.905	56 Ba barium 137.34	57-71 ○ rare earth metals	72 Hf hafnium 178.49	73 Ta tantalum 180.948	74 W tungsten 183.85	75 Re rhenium 186.20	76 Os osmium 190.2	77 Ir iridium 192.2
87 Fr francium 223	88 Ra radium 226.05	89-103 □ Ac actinium					
		57 ○ La lanthanum 138.91	58 Ce cerium 140.12	59 Pr praseo dymium- 140.907	60 Nd neodym- ium 144.24	61 Pm prom- ethium 145	62 Sm samarium 150.35	63 Eu europ 151.96
		89 □ Ac actinium 227	90 Th thorium 232.038	91 Pa proto- actinium 231	92 U uranium 238.03	93 Np neptunium 237	94 Pu plutonium 244	95 Am ameri 243

		non-metals					inert gases
third transition metals		boron and carbon groups		nitrogen and oxygen groups		the halogens	
group 1b	group 2b	group 3a	group 4a	group 5a	group 6a	group 7a	group 0
							2 He helium 4.003
		5 B boron 10.81	6 C carbon 12.011	7 N nitrogen 14.0067	8 O oxygen 15.9994	9 F fluorine 18.9984	10 Ne neon 20.179
		13 Al aluminium 26.9815	14 Si silicon 28.086	15 P phos-phorus 30.9738	16 S sulphur 32.064	17 Cl chlorine 35.453	18 Ar argon 39.948
29 Cu copper 63.546	30 Zn zinc 65.37	31 Ga gallium 69.72	32 Ge germanium 72.59	33 As arsenic 74.9216	34 Se selenium 78.96	35 Br bromine 79.904	36 Kr krypton 83.80
47 Ag silver 107.868	48 Cd cadmium 112.40	49 In indium 114.82	50 Sn tin 118.69	51 Sb antimony 121.75	52 Te tellurium 127.60	53 I iodine 126.9044	54 Xe xenon 131.30
79 Au gold 196.967	80 Hg mercury 200.59	81 Tl thallium 204.37	82 Pb lead 207.19	83 Bi bismuth 208.98	84 Po polonium 209	85 At astatine 210	86 Rn radon 222

65 Tb terbium 158.94	66 Dy dysprosium 162.50	67 Ho holmium 164.90	68 Er erbium 167.26	69 Tm thulium 168.934	70 Yb ytterbium 173.04	71 Lu lutetium 174.97
97 Bk berkelium 247	98 Cf californium 251	99 Es einsteinium 254	100 Fm fermium 257	101 Md mendelev-ium 258	102 No nobelium 255	103 Lw lawrencium 257

The 19th-century chemists thought that atoms were held together by chemical affinity, or even by little hooks. However, as the structure of atoms became clear at the beginning of this century, it was evident that the bonds between atoms are basically due to electrical attraction. The model of the atom that has evolved is of a central positively charged nucleus surrounded by one or more negatively charged electrons. These electrons are arranged in groups (shells) around the nucleus. A study of the periodic table and the pattern of the electron shells makes it clear that chemical properties depend almost entirely on the number of electrons in the outer shell of an atom.

The first electron shell holds only two electrons. The second shell holds up to eight electrons. The third shell fills up to a stable arrangement with eight electrons. The shells then divide into sub-shells and the arrangement becomes more complex, but the outer shell, or sub-shell, never holds more than eight electrons. An outer shell, or sub-shell of eight electrons, or two electrons for the first shell, is a very stable arrangement and elements having those numbers of electrons do not form normal compounds; they are the inert gases helium (two

the first 18 elements of the periodic table

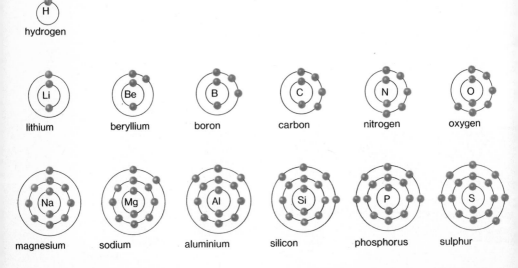

electrons), neon (2·8 electrons) and argon (2·8·8 electrons).

When atoms combine together they strive to form more stable electron shell arrangements, similar to the inert gases. For example, the element sodium has one electron in its outer shell, and chlorine has seven electrons. When these two elements combine to form sodium chloride (common salt) there is a transfer of one electron from the sodium atom to the chlorine atom producing a stable compound, each atom ending up with eight electrons in its outer shell. This is known as electrovalent bonding because the transfer of the electron makes the sodium atom positively charged and the chlorine atom negatively charged. The charged atoms (ions) attract each other and form a rigid crystal.

Another method of combination is by sharing electrons in covalent bonding. Oxygen with six electrons in its outer shell shares the single electrons of two hydrogen atoms to make a complete eight-electron shell. These shared electron-pairs also complete the first electron shells of the hydrogen atoms. All compounds are formed by one of these methods or sometimes by a combination of both.

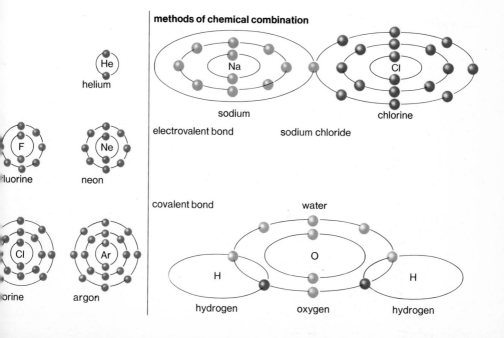

methods of chemical combination

He
helium

F
fluorine

Ne
neon

Cl
orine

Ar
argon

Na
sodium

Cl
chlorine

electrovalent bond   sodium chloride

covalent bond   water

H
hydrogen

O
oxygen

H
hydrogen

More than half the Earth's surface is covered with water, making it the most common substance. To the early Greeks water was one of the four 'elements' – Earth, water, air and fire – from which all other materials were made. Water was shown to be a compound of the gases hydrogen and oxygen by the English scientist, Henry Cavendish, in 1766. The simplest way of showing that water is a compound substance is by passing an electric current through it. If two wires from a battery are dipped into water, bubbles of hydrogen gas form at the wire attached to the negative terminal and bubbles of oxygen at the wire attached to the positive terminal. If these two gases are collected it will be found that the volume of hydrogen formed is twice as great as the volume of oxygen. As equal volumes of gas contain equal numbers of atoms or molecules (Avogadro's hypothesis), it follows that the formula for water must be H_2O.

Water is a solvent for many chemicals. Natural spring water formed underground when rain runs over rocks contains dissolved materials. Hard water, in which it is difficult to make a lather with soap, contains dissolved calcium and magnesium salts. They combine with the soap

Right: hard water running through a pipe leaves a deposit of calcium and magnesium salts

electrolysis

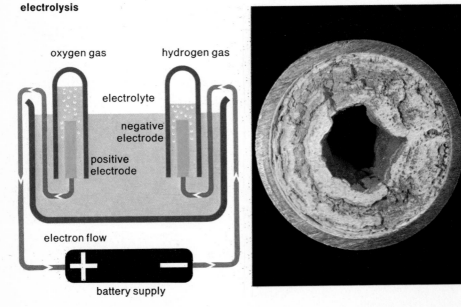

oxygen gas

hydrogen gas

electrolyte

negative electrode

positive electrode

electron flow

battery supply

to make an insoluble scum. One advantage of detergents over soaps is that a smaller quantity is required in hard-water districts.

Many crystalline substances contain combined water (water of crystallization). When heated, the crystals lose this water and turn into powder. Blue crystals of copper sulphate turn into a white powder when the water of crystallization is removed. Common crystals which do not contain water of crystallization are salt, sugar, quartz crystals (sand) and diamonds. They are anhydrous (without water).

As a gas (steam) water consists of separate molecules comprising an oxygen atom bonded to two hydrogen atoms, with an angle of about 105° between the bonds. In the liquid state the individual molecules are held together by hydrogen bonds. These are relatively weak bonds that form between the oxygen atoms of one molecule and the hydrogen bonds of another. In solid water (ice) the hydrogen bonds hold the water molecules in an open rigid structure. When the ice crystals melt some of the hydrogen bonds break and the individual water molecules come closer together. This is why ice is lighter than water which has its maximum density at 4°C.

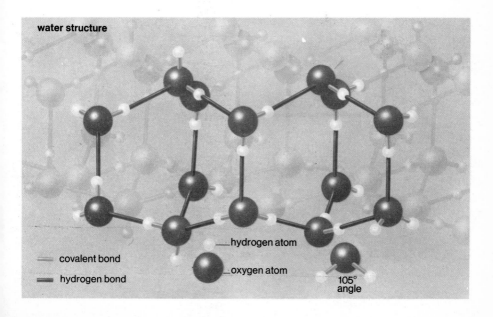

water structure

covalent bond

hydrogen bond

hydrogen atom

oxygen atom

105° angle

Chemistry
141 analysis and synthesis

Two of the main procedures carried out by chemists are analysis and synthesis. Analysis is finding out exactly what a substance consists of in terms of mixtures, compounds and elements. Once that is determined a chemist can frequently make the substance himself in the laboratory. Synthesis is the making of complex materials from simpler ones, and man-made materials are often called synthetic.

In both analysis and synthesis it is often necessary to separate solids from liquids. Some solid substances seem to disappear when they are mixed with a liquid. They dissolve to form a solution, as when salt dissolves in water to form a salt solution. Substances which do not dissolve in a particular liquid are said to be insoluble in that liquid. An insoluble substance can be separated from a soluble one by filtering. On a small scale this is done by pouring the solution through a filter-paper into a filter-funnel. The insoluble substance remains on the filter-paper while the liquid (filtrate) passes through it. On an industrial scale, a filter-press or rotary filter is used.

A dissolved substance (solute) can be separated from the liquid in which it is dissolved (solvent) by boiling away the solvent. This process

evaporation

liquid boiled away,
solid substance remains

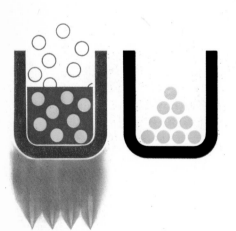

distillation

liquid boiled and gas cooled,
liquid condenses

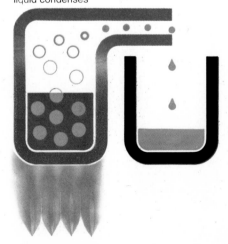

is evaporation. On the other hand, the solvent can be separated by distillation. This involves boiling the solution and cooling the vapour given off so that it condenses back to a liquid. Distillation is a very ancient process which the alchemists used for making perfumes and spirits. Two or more liquids mixed together can be separated by fractional distillation, provided that they boil at different temperatures. The liquid that boils at the lowest temperature is driven off first, then the next, and so on. It is by this process that petrol, paraffin and lubricating oils are separated from crude oil in refineries.

Very similar substances that are difficult to separate chemically can sometimes be isolated by using chromatography, a form of analysis. A solution of the mixed substances is poured through a column of absorbent material, or a piece of absorbent paper is dipped in the solution. The different substances are absorbed from the solution at different rates and so form a series of coloured bands. Originally the method could only be used for materials of different colours but now colourless materials can be identified by staining them or by fluorescence with ultraviolet light.

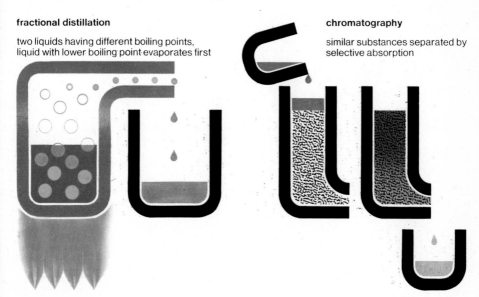

fractional distillation

two liquids having different boiling points, liquid with lower boiling point evaporates first

chromatography

similar substances separated by selective absorption

One of the most important methods used in science is to classify materials with similar properties. For example, the many substances found to be corrosive and with a bitter taste were called acids by the early chemists. Other substances, which could neutralize the effects of these acids, were named alkalis. These group names have been retained but today they have much more precise meanings.

Most acids react with metals, giving off hydrogen gas, but all acids form hydrogen ions (H^+) when dissolved in water. Acids are, in fact, electrovalent compounds of hydrogen and other atoms or groups of atoms called radicals. The three common mineral acids are: sulphuric acid, H_2SO_4, which in solution in water consists of $2H^+ + SO_4^{2-}$; hydrochloric acid, HCl ($H^+ + Cl^-$) and nitric acid, HNO_3 ($H^+ + NO_3^-$). The SO_4^{2-} group of atoms is the sulphate radical – a covalent group of one sulphur atom and four oxygen atoms with two extra electrons to give it the two negative charges. Similarly the NO_3^- group is called the nitrate radical.

A commonly occurring radical is OH^- (the hydroxyl group), a covalent group of one oxygen atom and one hydrogen atom. This

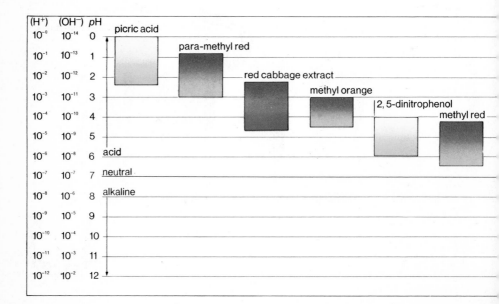

will readily combine electrovalently with a metal, or sometimes with other groups of atoms, to form an alkali. When dissolved in water, alkalis produce hydroxyl ions, OH^-. For example, sodium hydroxide (caustic soda), NaOH, in solution consists of $Na^+ + OH^-$, and ammonium hydroxide, NH_4OH, becomes $NH_4^+ + OH^-$. A base includes alkalis and some insoluble oxides which are also able to neutralize acids.

The result of neutralizing an acid with an alkali is always to form a salt and water; the hydrogen ions of the acid and the hydroxyl ions of the alkali combine covalently to form water.

Since acids, alkalis and salts are all electrovalent compounds they all produce ions when dissolved in water and these charged particles enable them to conduct electricity. Acids, alkalis and salts are all called electrolytes.

Some dyes, known as indicators, change colour in the presence of an acid or alkali. The colour depends on the concentration of hydrogen ions or hydroxyl ions in the solution. The most common is litmus which turns red with acids and blue with alkalis.

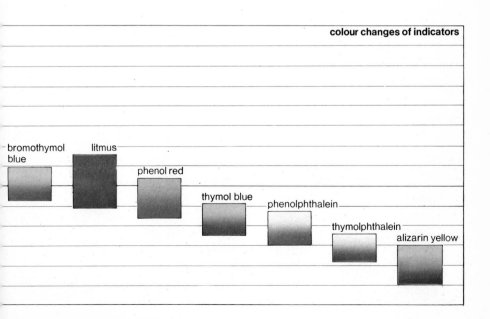

colour changes of indicators

bromothymol blue · litmus · phenol red · thymol blue · phenolphthalein · thymolphthalein · alizarin yellow

Common salt (sodium chloride) is a compound of sodium and chlorine ions arranged in a regular cubic lattice. Each crystal of the salt we use in food consists of a cubic sodium-chlorine lattice repeated millions and millions of times. Salt is essential to human life because our blood contains about one-half percent of it. In fact, some people believe that animals originated in the sea because human blood contains approximately the same proportion of salt as sea water. When we perspire, cry or urinate, the salt loss must be replaced from our diet.

Salt is obtained from the sea and in the form of rock salt in places such as Britain, Poland and the United States. The chemical industry uses over thirty million tons of salt a year. In Poland salt is mined in the same way as coal. In parts of Britain the salt is dissolved in water in the mine, the brine is pumped to the surface and then the water is evaporated, leaving the solid salt to crystallize out.

Salt is the starting material for the manufacture of all chemicals that contain either sodium or chlorine, and more chemicals in everyday use are compounds of sodium than compounds of any other element. The most widely used process for manufacturing chemicals

Photomicrograph of salt crystals magnified 145 times

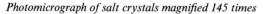

from salt is electrolysis. This consists of passing a current of electricity through brine so that the sodium chloride is split into sodium and chlorine ions. The sodium reacts with the water to form a solution of caustic soda (sodium hydroxide, NaOH) which is the basis for the manufacture of all the sodium chemicals. Chlorine and hydrogen are also produced as by-products. In the early days of the chemical industry, the chlorine by-product was considered a nuisance because it is a poisonous green gas and is difficult to dispose of in safety. Today there are so many new compounds containing chlorine that it is even more important than sodium hydroxide.

Apart from its use in hydrochloric acid, chlorine is widely used in water purification to kill bacteria. Many antiseptics, disinfectants, insecticides, drugs and anaesthetics (such as chloroform) contain chlorine. It is now extensively used in the manufacture of the plastic PVC (polyvinylchloride). Raincoats, floor coverings, electrical insulation, records and many other household objects are made of this tough, flameproof plastic. More than two hundred thousand tons of chlorine a year are produced in the United Kingdom alone.

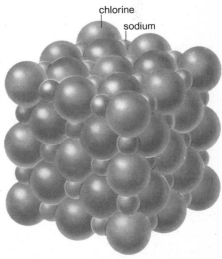

a cubic crystal of sodium chloride (salt)

chlorine

sodium

Sulphur, like salt, is an important mineral because it is the basis for the manufacture of many essential materials. Most industrial sulphur is in the form of sulphuric acid. Over thirty million tons a year are produced and it has been called the world's most important chemical. Sulphur is found underground, combined with metals, as sulphides of iron, silver, lead, copper and zinc. There are deposits of crude sulphur in Sicily and in Texas and Louisiana in the United States; these supply 95% of the world's sulphur. One way of extracting sulphur from the ground is the Frasch process.

The formula for sulphuric acid is H_2SO_4; if this is written in the form $H_2O.SO_3$ it shows that it is a compound of water, H_2O, and sulphur trioxide, SO_3. When sulphur burns in air it forms sulphur dioxide, SO_2, a choking poisonous gas. To create sulphuric acid, the industrial chemist combines the sulphur dioxide with another atom of oxygen to produce sulphur trioxide. This is a difficult process and it needs the aid of a catalyst to make it commercially possible. A catalyst is a substance which alters the rate of a chemical change while the catalyst itself is unaltered at the end of the reaction.

Finely divided (ground) platinum was first used as a catalyst in the sulphur trioxide process but has now been replaced by vanadium pentoxide. In the contact process the sulphur dioxide and oxygen are compressed and passed over the heated catalyst so that they combine to form sulphur trioxide, $2SO_2 + O_2 = 2SO_3$. The sulphur trioxide is then dissolved in water to form sulphuric acid.

Catalysts are very important in the chemical industry. In the manufacture of synthetic ammonia, NH_3, from nitrogen and hydrogen, the heated gases are made to combine under high pressure with the help of finely divided osmium as a catalyst. Ammonia is used in the manufacture of fertilizers and explosives.

Sulphur dioxide is produced during the burning of coal and oil. It is estimated that six million tons per year of sulphur dioxide are discharged into the atmosphere of Great Britain alone, causing great damage to buildings and health. The Clean Air Act of 1956 greatly reduced the formation of thick fogs in British towns by creating smokeless zones and limiting air pollution.

Frasch process

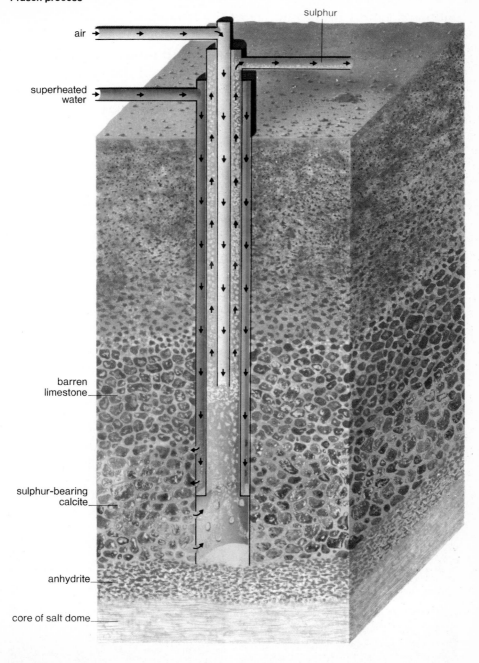

sulphur

air

superheated
water

barren
limestone

sulphur-bearing
calcite

anhydrite

core of salt dome

It is hard to imagine modern industrial life without an almost unlimited supply of iron and steel for buildings, cars, ships, trains and countless vital machines. But metals such as iron are rarely found underground in a usable form because they are usually combined with other elements into compounds. The mixture of the chemical compound (mineral) with the rocky material in which it is found is an ore.

The common ores of iron are haematite (iron oxide), iron pyrites (iron sulphide) and siderite (iron carbonate). To obtain the pure metal from the ore, two problems have to be solved. First the impurities must be removed to separate the minerals from the ore, and then the metal has to be extracted from its compound. Much of the earthy material can be washed away from the heavier ore. The ore is then roasted in air and the iron, if it is not already present as an oxide, combines with oxygen to form iron oxide. The iron oxide is then mixed with carbon (coke) which combines more easily with the oxygen than it does with the iron. The removal of oxygen from a compound is known chemically as reduction, and the combined process of roasting and reduction is smelting. Starting with iron pyrites, FeS_2,

In a steel mill, molten steel is sampled (1) before being poured into ingot moulds (2). After the steel ingots are cooled, they are transferred to the soaking pits (3) to be heated to a temperature suitable for rolling. The rolled slabs are reduced to thin sheet steel (4) and sold as cold steel coils (5).

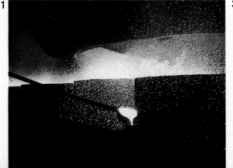

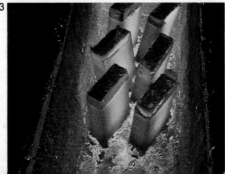

the smelting processes are summarized in the equations:

$$4FeS_2 + 11O_2 = 8SO_2 + 2Fe_2O_3$$
$$2Fe_2O_3 + 3C = 3CO_2 + 4Fe$$

The process of extracting iron is usually carried out in a blast furnace in which the coke is ignited at the bottom of the furnace by a blast of hot air, producing a temperature high enough (over 1500°C) to melt the iron. Limestone is mixed with the ore and combines with the impurities in it to form a slag. The molten iron is run from the furnace into moulds and is then known as pig iron; it contains about 5% impurities, mainly carbon. When used in this form it is cast iron. A disadvantage of iron is that it slowly recombines with the oxygen in the air when it rusts. Painting prevents rusting, or the iron can be plated with another metal such as zinc (galvanized iron). Steel contains between 1–2% carbon and is much stronger than cast iron.

A mixture of two or more metals is an alloy. Stainless steels are alloys of steel and chromium and do not rust. They usually contain 70–90% iron, 12–20% chromium and 0·1–0·7% carbon.

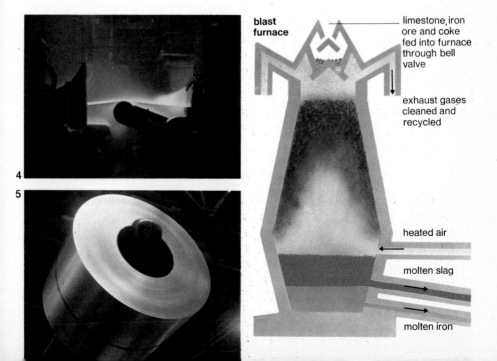

4
5

blast furnace

limestone, iron ore and coke fed into furnace through bell valve

exhaust gases cleaned and recycled

heated air

molten slag

molten iron

Some common metals (for example, copper, tin, zinc and lead) can be extracted from their ores by a reduction process with carbon in the same way as the smelting of iron. In fact, copper was one of the earliest metals extracted by man. It was probably discovered when copper ores were accidentally heated in a fire and reduced by the carbon present in the charcoal.

Although copper appears to be a very common metal, it is relatively rare – it makes up only 0·01% of the Earth's crust. Aluminium, on the other hand, is the third most common element (about 8%) and is in more abundant supply than any other metal. However, it was among the last of the common metals in use today to be extracted from its ore. This is because it combines very firmly with oxygen and the aluminium oxide cannot be reduced with carbon. Ordinary garden clay contains some 25% aluminium combined with other elements, such as silicon, but the aluminium cannot be easily extracted.

The principal commercial ore of aluminium is bauxite (impure aluminium oxide). The ore is first purified chemically with caustic soda solution in which the aluminium oxide dissolves, allowing the

electrolytic production of aluminium

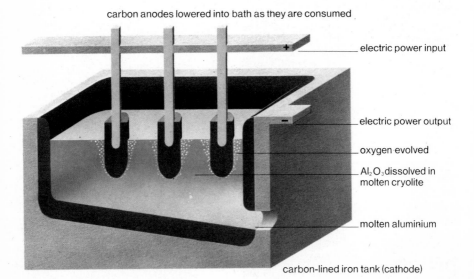

carbon anodes lowered into bath as they are consumed

electric power input

electric power output

oxygen evolved

Al_2O_3 dissolved in molten cryolite

molten aluminium

carbon-lined iron tank (cathode)

impurities to be filtered off. The solution is heated so that when the water has evaporated pure aluminium oxide (alumina) remains. The alumina is then split into oxygen and aluminium by electrolysis. This is done by dissolving the alumina in molten cryolite (natural sodium aluminium fluoride) at 1000°C and passing an electric current through it. The aluminium is set free at the negative electrode and oxygen is given off at the positive electrode. A large electric current is necessary and it takes some 20 000 kWh of electricity to produce a ton of aluminium. It is only economical to produce aluminium where electricity is cheap – usually in places where hydroelectric power is readily available.

Aluminium is useful because it is very light, is a good conductor of heat and electricity, and it forms alloys which are also light and strong. It does not rust because the very thin film of aluminium oxide which forms on the surface protects it from further corrosion. Alloys often possess quite different properties from the metals from which they were made, and by adding different amounts of different metals the metallurgist can produce alloys with special properties.

A factory in Scandinavia which manufactures aluminium

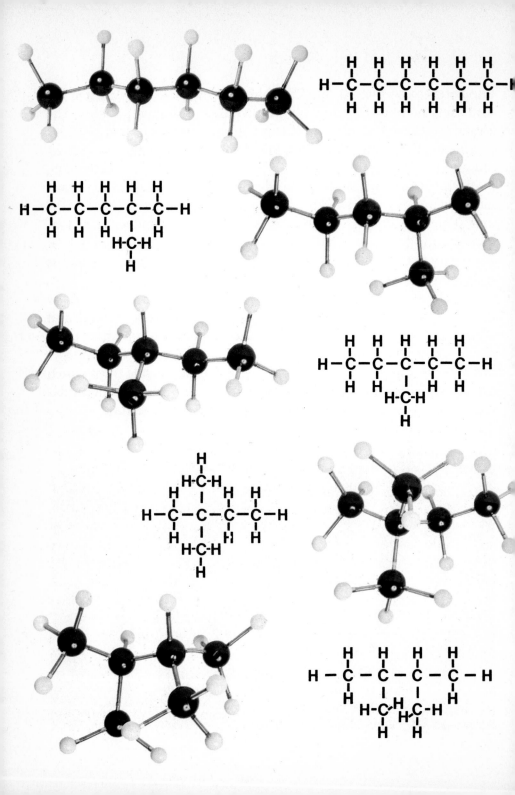

The oils, sugars, resins and fats found in plants and animals are called organic substances because they are produced by living organisms. Substances produced from minerals are inorganic (non-living). Until well into the 19th century it was thought that the organic chemical changes taking place in living things differed in some way from the inorganic chemical changes carried out in the laboratory. Now it is accepted that the laws governing chemical changes are the same whether the reaction takes place in a test tube or in a living cell.

Organic chemistry now means the chemistry of carbon compounds because the great majority of substances produced during living processes contains the element carbon. There are more compounds containing carbon than there are compounds of all the other elements put together. Most of these are compounds of carbon and hydrogen, sometimes with oxygen and nitrogen included. The main reason for the great diversity of carbon compounds is that carbon atoms have the unique property of being able to combine with themselves to form straight-chain molecules and ring-shaped molecules. Silicon can do this to a lesser extent. The straight-chain molecules are aliphatic compounds and include the paraffins, olefines, acetylenes, alcohols and fatty acids. The ring-shaped molecules are aromatic compounds and are based on benzene produced from coal tar.

Another reason for the great number of carbon compounds is that the carbon atoms in a compound can be arranged in many different ways, each of which forms a different type of molecule. Two or more compounds each of which consists of the same atoms arranged differently are isomers (*iso* = equal, *mer* = part). The larger the molecule the greater the number of ways the atoms can be arranged. The atoms of a simple hydrocarbon such as hexane, C_6H_{14}, which is present in petrol can be arranged in five different ways forming five different types of molecules, thus making five isomers of hexane.

Most of the synthetic organic chemicals in daily use such as drugs, dyes, plastics and synthetic fibres are derived from oil or coal which are themselves the products of fossilized animal and vegetable products. In this way there is a continuing association of organic chemistry with the products of living organisms.

Aliphatic compounds have straight-chain molecules in which the carbon atoms are linked together by covalent bonds. There are four electrons in the outer electron shell of the carbon atom, so that by combining with four atoms of hydrogen complete electron shells are produced and the compound methane, CH_4, is formed. This is the first member of the paraffin series in which successive members are formed by replacing a hydrogen atom with another carbon atom to which two hydrogen atoms are attached: ethane, C_2H_6, propane, C_3H_8, butane, C_4H_{10} and octane, C_8H_{18}. The general formula for the series, which is sometimes called the alkane series, is C_nH_{2n+2}. The paraffins exist naturally in oil and natural gas.

The hydrogen atoms in the paraffins can be replaced with other types of atoms, such as the halogens (chlorine, iodine, bromine and fluorine) to form more substitution compounds. For example, if three of the hydrogen atoms of methane are replaced with chlorine atoms, trichloromethane, $CHCl_3$, or chloroform, is formed.

Two carbon atoms can link together using two shared electron pairs – this is known as a double bond. A series formed in this way is

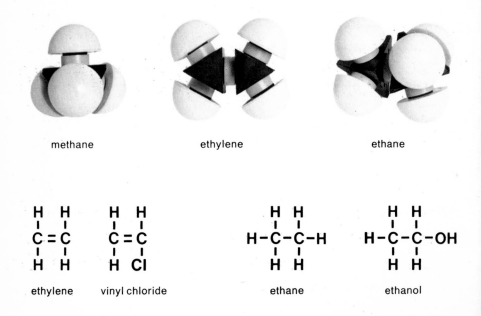

methane   ethylene   ethane

ethylene   vinyl chloride   ethane   ethanol

the olefines, or alkenes, with the general formula C_nH_{2n}. These compounds are said to be unsaturated because if the double bond is replaced by two single bonds another two hydrogen atoms can be added to their molecules. The paraffins are saturated compounds: there are no extra bonds available. The hydrogen atoms of olefines can also be replaced by halogen atoms. Thus ethylene, $CH_2{=}CH_2$, the first member of the olefine series, can produce vinyl chloride, $CH_2{=}CH.Cl$; when many of these molecules link together the plastic polyvinylchloride or PVC is formed.

When two carbon atoms link together using three shared electron pairs there is a triple covalent bond. A series built up in this way is the acetylenes, or alkynes, with the general formula C_nH_{2n-2}. The most important member of this group is the gas acetylene, $HC{\equiv}CH$.

The hydroxyl group of atoms, $-OH$, can be substituted for one of the hydrogen atoms in an organic compound and the result is another series called the alcohols. Methanol (methyl alcohol), CH_3OH, comes from methane, ethanol (ethyl alcohol), C_2H_5OH, comes from ethane and so on.

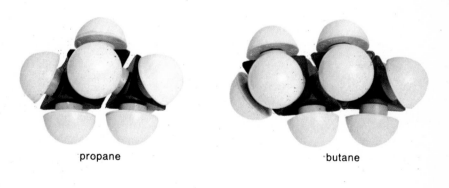

propane   ·butane

methane   chloroform   methanol   acetylene

The word aromatic means pleasant-smelling, and was first applied to certain naturally occurring oils and essences to distinguish them from the paraffin hydrocarbons. Aromatic compounds contain one or more benzene rings. Benzene, discovered by the British scientist, Michael Faraday, in 1825, is found in the coal tar produced when coal is distilled in making coal gas. The formula for benzene is C_6H_6 and it was first shown by Friedrich Kekulé, the German chemist, that the six carbon atoms are joined together to form a ring with alternate single and double bonds.

Benzene, a colourless liquid, is a good solvent for many organic substances, but its real importance is in the large number of substitution products that can be obtained from it. The hydrogen atoms can be replaced by other single atoms or groups of atoms.

The replacement of hydrogen atoms by hydroxyl groups produces new compounds called phenols. Phenols, unlike aliphatic compounds containing —OH groups, are acids, not alcohols. This is because the hydrogen of the —OH group is linked electrovalently to the benzene ring and not covalently as in alcohols. Phenol, also known as carbolic

benzene ring

aniline

phenol

acid, is $C_6H_5 . OH$ and was used by the British surgeon, Joseph Lister, as an antiseptic in 1865. Many modern disinfectants, antiseptics and plastics are phenol derivatives.

Aniline, $C_6H_5 . NH_2$, is produced by substituting an NH_2 group (the amino group) for one of the hydrogen atoms in the benzene ring. Aniline is the basis of aniline dyes and many drugs including the sulphanilamides. Another important derivative is toluene, $C_6H_5 . CH_3$, in which a methyl group replaces a hydrogen atom. Replacing a further three hydrogen atoms in the benzene ring by nitro groups, NO_2, produces the explosive trinitrotoluene (T.N.T.), $(NO_2)_3 . C_6H_2 . CH_3$.

As there are six points in the benzene ring at which substitutions can be made, the number of aromatic compounds is very large. In fact, the complexity of the substitutions can vary considerably from a single atom to a long 'side chain' and two or more rings can be joined together in a number of ways. More than a thousand different dyes, drugs, paints, solvents, scents, essences, adhesives, plastics and synthetics are derived from coal tar.

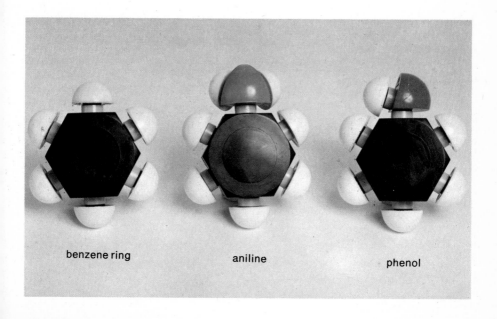

benzene ring   aniline   phenol

The molecular formula of a chemical compound shows its atomic make-up. The formula for aspirin, or acetylsalicylic acid, is $C_9H_8O_4$. This tells us the composition of aspirin and, by using the atomic weights, the relative proportions by weight of the elements in it. One hundred and eighty grams of aspirin contain $9 \times 12 = 108$ grams of carbon (9 carbon atoms each with an atomic weight of 12), $8 \times 1 = 8$ grams of hydrogen and $4 \times 16 = 64$ grams of oxygen. But the formula does not show how the atoms are arranged in the molecule. The molecular pattern is given by the structural formula.

Chemists can work out the pattern of molecules by splitting them into smaller groups which can be identified. With this additional information aspirin can be written $C_6H_4 . CO_2CH_3 . COOH$, showing that it is derived from benzene, C_6H_6, and acetic acid, $CH_3 . COOH$. Acetic acid is a derivative of methane, in which one hydrogen atom is replaced by the fatty acid (carboxyl) group $-COOH$. Finally the exact arrangement of the atoms can be worked out. The structural formula must be known if the compound is to be made synthetically.

In crystalline substances the shape of the crystal can give some clue

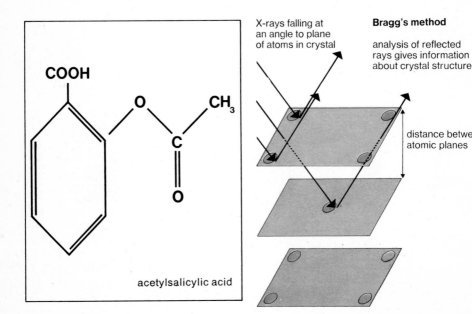

COOH

O   CH₃

C

O

acetylsalicylic acid

X-rays falling at an angle to plane of atoms in crystal

Bragg's method

analysis of reflected rays gives information about crystal structure

distance between atomic planes

to the atomic arrangement. Crystals are solids with a definite shape resulting from the regular arrangement of the atoms. In a salt crystal the atoms are arranged at the corners of cubes. In other crystals there is an extra atom at the centre of the cube (body-centred crystal) or at the centre of each cube face (face-centred crystal).

The shape of very large molecules can be photographed with an electron microscope but a more precise method of studying molecular arrangement is from X-ray diffraction patterns. When a beam of X-rays passes through a crystal the paths of the X-rays are reflected by the atoms they meet. The pattern produced on a photographic plate by the reflected rays can be used to work out the atomic arrangement. This method was invented in 1913 by the English physicists, Sir William Bragg, and his son, Sir Lawrence Bragg. It is complicated because a pattern has to be worked out in three dimensions, but it has led to very important results. The structure of the very complex molecule of DNA (deoxyribonucleic acid), the substance concerned with the transmission of hereditary factors in living cells, was worked out in this way. So, too, were some of the simpler proteins.

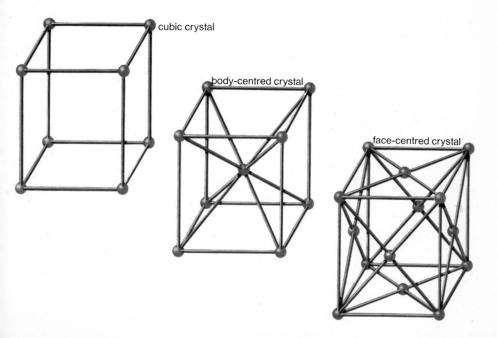

cubic crystal

body-centred crystal

face-centred crystal

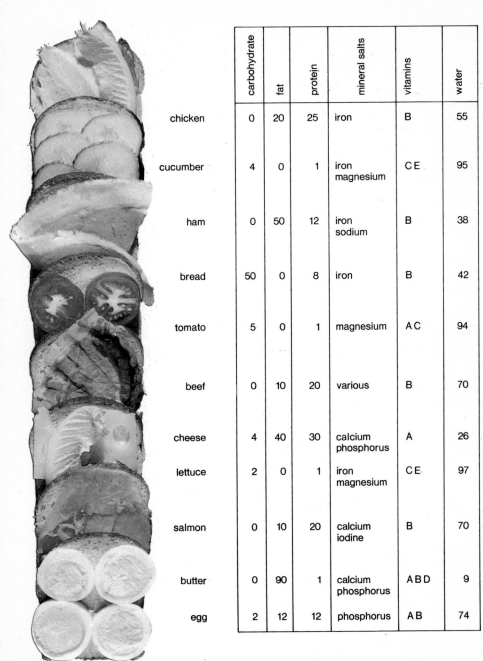

	carbohydrate	fat	protein	mineral salts	vitamins	water
chicken	0	20	25	iron	B	55
cucumber	4	0	1	iron magnesium	C E	95
ham	0	50	12	iron sodium	B	38
bread	50	0	8	iron	B	42
tomato	5	0	1	magnesium	A C	94
beef	0	10	20	various	B	70
cheese	4	40	30	calcium phosphorus	A	26
lettuce	2	0	1	iron magnesium	C E	97
salmon	0	10	20	calcium iodine	B	70
butter	0	90	1	calcium phosphorus	A B D	9
egg	2	12	12	phosphorus	A B	74

contents in percentages

All living things must have food to provide new material for growth and to supply energy. The chemical changes that take place in living cells are its metabolism. In some of these changes large food molecules are broken down into simpler ones (catabolism) with the release of energy, as in digestion and respiration. In other changes complex molecules for cell construction and function are built up (anabolism), as in photosynthesis.

All of our food consists primarily of only a few types of carbon compounds. These are carbohydrates, fats and proteins which together with vitamins, mineral salts and water make up our diet.

Carbohydrates are compounds of carbon, hydrogen and oxygen with the general formula $C_n(H_2O)_m$, where n and m can have different values. In pentose sugars, such as ribose, $C_5H_{10}O_5$, $n = 5$; in hexose sugars such as glucose, $C_6H_{12}O_6$, $n = 6$. In both cases the sugars are monosaccharides as they consist of only one saccharide unit. However, glucose units can link together to form larger units; for example, sucrose is a disaccharide (two units), $C_{12}H_{22}O_{11}$. Starch is a polysaccharide in which hundreds of glucose units link together, while cellulose is a form of starch in which there are strong cross-linkages between the glucose units. Sugar is produced in sugar cane and sugar-beet and starch is stored in seeds, roots and tubers.

Fats are carbon, hydrogen and oxygen compounds but the proportion of oxygen is less than in the carbohydrates. Chemically, fats and oils are compounds of glycerol (a form of alcohol with three —OH groups) with fatty acids. Butter, lard and suet are animal fats. Liquid fats or oils can be obtained from plants – olive oil, coconut oil and peanut oil. Fats produce energy and heat when oxidized, but because they contain a greater proportion of hydrogen and carbon the energy output is greater than in carbohydrates.

Vitamins are substances needed in very small quantities for the proper growth and maintenance of the body. They are organic chemicals and vary from comparatively simple molecules like vitamin C (ascorbic acid), $C_6H_8O_6$, to complex molecules like vitamin B_{12}, $C_{63}H_{90}O_{14}N_{14}PCo$, the absence of which produces pernicious anaemia.

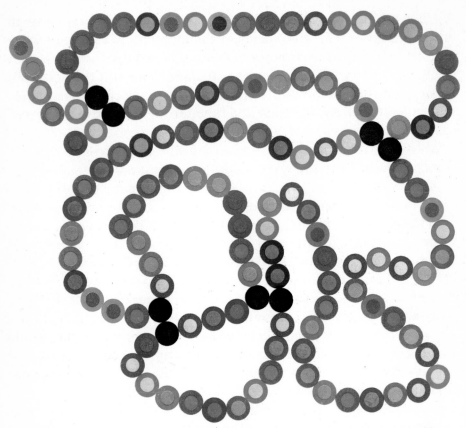

sequence of amino acids and shape of the protein lysozyme which occurs in raw egg white

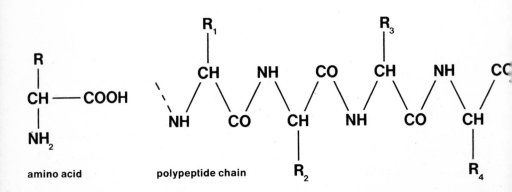

amino acid

polypeptide chain

Cheese, lean meat, chicken, eggs, fish, nuts, peas, beans and flour are rich in proteins. Not only is a supply of protein needed for growth and the replacement of worn-out tissues, but many of the essential chemicals of the body are made from the units of which protein molecules are constructed. Proteins consist of large and complex molecules containing carbon, hydrogen, oxygen and nitrogen atoms and sometimes also sulphur and phosphorus. They are a most important part of our diet because they are essential constituents of all living things from the simplest plant to man himself.

Proteins are built from amino acid units. These are carbon and hydrogen compounds containing both carboxyl, $-COOH$, and amino, $-NH_2$, groups. The carboxyl group of one amino acid can join with the amino group of another amino acid to form very large molecules. The linkage between the amino acids is known as a peptide linkage, and thus the long-chain protein molecules are poly-peptides. There are 21 amino acids occurring in nature, thus allowing an enormous number of ways in which they can link together to form different proteins: without this great diversity in proteins, life would not be possible. This is why protein chemistry is so complicated. The first protein to have its structure worked out was the hormone insulin, which controls the body's sugar metabolism. In 1952 Frederick Sanger, a British biochemist, found that it consists of four separate polypeptide chains linked together, two chains having 21 amino acids and two chains having 30 amino acids. When proteins are digested they are broken down into their constituent amino acids, which are absorbed into the blood and reassembled in the body to form new proteins.

One of the most important groups of proteins in living things are the enzymes. These are biochemical catalysts and all the reactions that take place in living cells depend on them. Each stage in a series of reactions requires its own special enzyme to make it work, and there are therefore thousands of different enzymes in living organisms. Enzymes are especially important in living cells because their production is controlled by the genes that are handed down from generation to generation. Each gene is responsible for producing one enzyme.

Chemistry
153   chemicals from coal and oil

Coal is a fossil formed from the cellulose of vegetation that grew during the Carboniferous period 350 million years ago. A typical coal sample might consist of 85% carbon, 9% oxygen, 5% hydrogen and 1% nitrogen, sulphur and mineral matter. To burn coal as a fuel is very wasteful because the many valuable chemicals which can be made from it are lost. When coal is heated without any air being present four products can be obtained: (1) coal gas, a mixture of hydrogen, methane, carbon monoxide, nitrogen and ethylene, (2) ammoniacal liquor, consisting of water and ammonia, (3) coal tar, which contains pitch, benzene, phenol, toluene, naphthalene and anthracene – all ring compounds of carbon and hydrogen, and (4) a solid residue, coke. Because all four products are very valuable it is more sensible to burn the coke as a fuel in homes and factories and burn coal only in a gas works. This would also cause considerably less atmospheric pollution.

The substances in tar can be separated by distillation because they have different boiling points, and they form the bases for a whole series of chemicals. More than 2000 different chemicals are derived

man-made materials

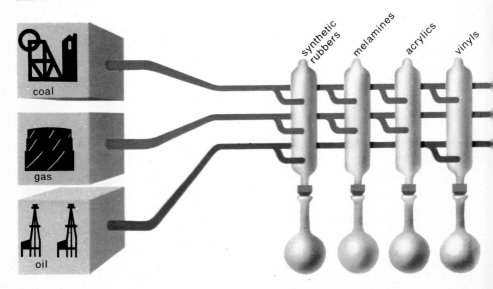

coal

gas

oil

synthetic rubbers

melamines

acrylics

vinyls

from coal. Benzene produces dyes, D.D.T., plastics and synthetic rubber. Phenol is used in antiseptics and in the manufacture of drugs, nylon and other plastics. Toluene is a basis for a whole range of chemicals from explosives to saccharine, naphthalene is used in dyes, resins and insecticides, and fertilizers are made with ammonia.

Oil is a mixture of hydrocarbons formed from the remains of prehistoric marine animals. The hydrocarbons are straight-chain compounds unlike the ring compounds of coal tar. The crude oil is separated in a refinery by fractional distillation into a range of products. Compounds with from six to 10 carbon atoms form the basis of motor fuels. However, ordinary fractional distillation produces many oils that are too heavy for use as petrol. These are further treated by heating to split the complicated larger molecules into smaller ones, which increases the output of petrol. This process, which requires a catalyst, is known as catalytic cracking and is the way that about half the world's petrol is made. Like coal tar, crude oil is the source of many organic chemicals and the petrochemical industry is now a very large part of the oil industry.

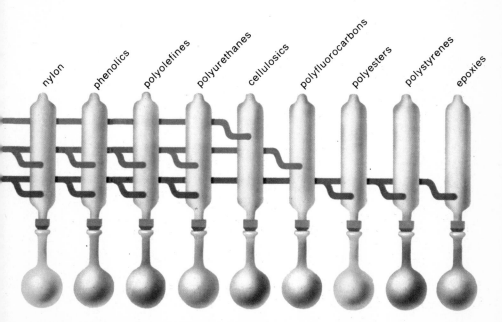

nylon phenolics polyolefines polyurethanes cellulosics polyfluorocarbons polyesters polystyrenes epoxies

Chemistry
154 plastics

Plastics are synthetic materials that can be softened and moulded into useful articles. Thermoplastics soften when heated and set hard when cooled and the process can be repeated again and again. Thermosetting plastics are moulded under heat and pressure during their formation but they then set hard and do not alter if heated again.

Plastics consist of very long-chain molecules which may contain many thousands of atoms. The long chains are formed by the joining together of large numbers of smaller molecules. The large molecules are polymers (*poly* = many, *mer* = a unit) and the simple molecules from which they are produced are called monomers.

Ethylene, C_2H_4, is a gas obtained from petroleum and natural gas. Its molecules contain six atoms and a double bond. When the gas is compressed to 2000 atmospheres at a temperature of about 200°C these small molecules join together. By a rearrangement of the links between the atoms a single long-chain molecule is formed containing some 2000 units and known to chemists as polyethylene. Polyethylene is a thermoplastic, easily moulded into thin sheets, bowls, bottles or pipes and is an excellent electrical insulator. A newer

phenol-formaldehyde plastic

type of thermoplastic is made from propylene, $CH_3CH{=}CH_2$, which polymerizes to form polypropylene. It is tougher than polyethylene and is used for sacks and carpet backing. A form of ethylene is the basis of fluorocarbons. The best known is tetrafluoroethylene, $CF_2{=}CF_2$, in which fluorine atoms replace the hydrogen atoms of ethylene. Polytetrafluoroethylene is heat-resistant and is used in non-stick cooking utensils.

In some polymerization processes two or more different monomer molecules combine by the elimination of water molecules between the joining atoms; this process is called condensation and the plastics formed are usually thermosetting compounds. Bakelite, one of the first plastics, results from the condensation of phenol and formaldehyde molecules. The long chains are held together by cross links in three dimensions and are very rigid. Another type of thermosetting resins are the epoxy resins made by condensing chloroepoxy propane, $CH_2.O.CH.CH_2Cl$, with phenols or glycols. They are used in fibreglass and adhesives.

Model of the structure of polythene plastic

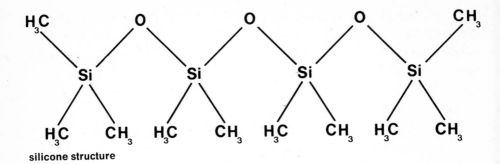

silicone structure

Natural rubber is made from a white milky liquid (latex) tapped from the bark of rubber trees. A Scotsman, Charles Macintosh, made raincoats with rubber in 1823 but when it was warm the crude rubber became soft and sticky. Rubber can be made less sticky by heating it with sulphur (vulcanizing). This process was developed by Charles Goodyear in the United States and patented in 1844. Rubber is tough, waterproof, and so elastic that it returns to its original shape when stretched or compressed. Materials that behave in this way are elastomers. Natural rubber is a polymer in which over ten thousand isoprene, $CH_2{=}C.CH_3.CH{=}CH_2$, units link together.

Synthetic rubber is made by polymerizing units with similar structural formulae to isoprene. Two of the substances used are butadiene, $CH_2{=}CH.CH{=}CH_2$, and chloroprene, $CH_2{=}C.Cl.CH{=}CH_2$. There are many others, however, and the properties of synthetic rubbers can be changed by altering the units from which they are made. Butadiene rubber is heat-proof, tough, and a good electrical insulator, and tyres, shoe soles, piping and flooring are made from it. Chloroprene rubber is very resistant to oils and solvents. Solid rocket fuel is manufactured from synthetic rubber (Thiokol) in which sulphur atoms replace some of the carbon atoms in the molecules.

Molecules, such as isoprene, that contain a double bond can be arranged in space in two ways: similar groups can be on the same side of the plane of the double bond (cis-form) or on the opposite sides (trans-form). This is called stereo-isomerism. Natural rubber consists exclusively of the cis-form of isoprene and by using special catalysts it is now possible to manufacture synthetic rubber composed of cis-poly-isoprene. These elastomers are stereo-regular rubbers.

Silicones do not occur naturally but are completely new synthetic molecules. Silicones consist of long-chain molecules in which the carbon atoms of plastics are replaced by alternate silicon and oxygen atoms. Silicones are water-repellent and make materials waterproof. They are used as lubricants for aircraft and guided missiles because they retain their properties at high and low temperatures. They are good electrical insulators and act as a bond between sheets of fibreglass to make a heat-resistant laminated material.

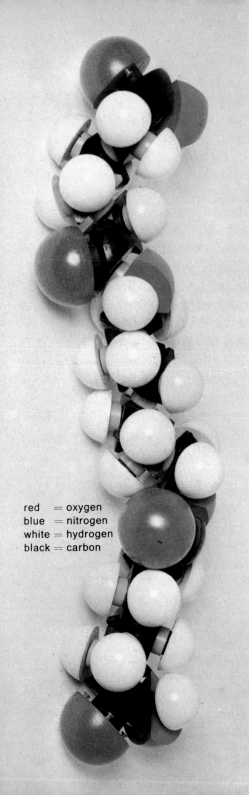

red = oxygen
blue = nitrogen
white = hydrogen
black = carbon

Man made one of his earliest discoveries when he began to farm. He found that he could spin the fibres of cotton, flax and wool into a yarn and then weave it into cloth. Cotton and flax, from which linen is made, are plant fibres while wool and silk are animal fibres. Fibres are long thread-like molecules. The structure of these large molecules and the way they are arranged in space determine the properties of the materials woven from them.

Wool and cotton are hard-wearing, good heat insulators and good moisture absorbers. But wool shrinks, creases and is attacked by moths. Cotton shrinks, crumples and is flammable. The early man-made fibres were imitations of natural fibres, but now that more is known about the dependence of properties on molecular structure, new fibres with new properties can be created.

Cotton is almost pure cellulose, which consists of cross-linked long chains of glucose units and is a glucose polymer. Rayon is manufactured by treating cellulose, obtained from wood pulp, with an alkali. Carbon disulphide

The molecular structure of nylon

is added and the syrupy liquid (viscose) formed is squirted through fine jets into sulphuric acid. The acid releases the cellulose from the compound it made with the alkali and fine threads of almost pure cellulose are formed. To make a non-inflammable artificial fibre, the cellulose is treated with acetic acid to form cellulose acetate. These fibres start with the naturally-occurring fibrous substance, cellulose.

Completely new fibres, which are not found in nature, are nylon, orlon and terylene. They are long-chain polymers using more complex units than cellulose. Nylon is made from adipic acid, $COOH(CH_2)_4COOH$, and diaminohexane, $NH_2(CH_2)_6NH_2$, both of which can be obtained from benzene from coal tar, or petroleum. These two molecules link together, eliminating water molecules, and form a chain with the structure $-NH-(CH_2)_6-$ $NH-CO-(CH_2)_4-CO-$. Nylon is strong and stretches. Creases set into terylene are long-lasting although the fabric itself does not crumple easily. The materials do not absorb water easily and thus 'drip-dry' quickly.

Nylon thread made by an extrusion process

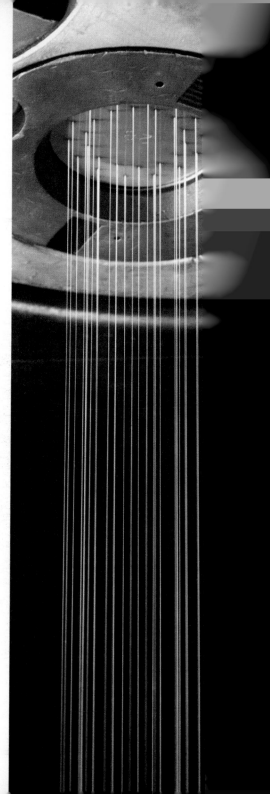

Like plastics and silicones, explosives are man-made. The first explosive was gunpowder – a mixture of charcoal, sulphur and potassium nitrate. It has been in use since the second half of the 13th century and for 600 years it was the only known explosive.

There are two principal requirements for an explosive: first, it must remain stable unless it is struck or ignited, and then, once it is ignited, a chemical reaction must take place to cause heat and a large volume of gas to be produced in a very short time. In gunpowder, the potassium nitrate, KNO_3, combines with the carbon in the charcoal to give the gas carbon dioxide, and also some nitrogen gas. A simplified equation of the reactions is: $2KNO_3 + S + 3C = K_2S + 3CO_2 + N_2$. Alfred Nobel, the Swedish founder of the peace prize, invented dynamite. It is nitroglycerin absorbed in kieselguhr, a natural substance made up of diatoms, the skeletons of minute plants. Nitroglycerin produces about ten thousand times its own volume of gas at 3000°C in a fraction of a second according to the equation $4C_3H_5(NO_3)_3 = 12CO_2 + 10H_2O + 6N_2 + O_2$.

toluene

trinitrotoluene (TNT)

glycerin (glycerol)

$$CH_2 - OH$$
$$CH - OH$$
$$CH_2 - OH$$

trinitroglycerin

$$CH_2 - NO_3$$
$$CH - NO_3$$
$$CH_2 - NO_3$$

Another explosive of this type is guncotton which is made of cellulose trinitrate. It is often mixed with nitroglycerin, which, as the equation shows, produces a surplus of oxygen. Cordite, the explosive used in shells and rifle bullets, is a mixture of 65% guncotton and 35% nitroglycerin. Another common explosive is T.N.T. (trinitrotoluene) but it contains too little oxygen in its molecule to produce carbon dioxide and instead forms carbon monoxide. T.N.T. is therefore usually mixed with ammonium nitrate which decomposes on heating to yield one atom of oxygen for every molecule of ammonium nitrate.

T.N.T. is very stable and requires a small explosion to set it off but a detonator must be used. Detonators are lead azide, PbN_6, and mercuric fulminate, $Hg(ONC)_2$. They decompose with explosive violence when struck with a hammer. The rifle or pistol trigger releases the hammer which strikes the detonator, which then explodes the cordite. As well as their uses in guns, explosives are also widely used for blasting and quarrying.

Explosives crumple a huge tower in seconds

Dyes are chemicals which colour fabrics. A satisfactory dye must attach itself firmly to the fabric so that it cannot be washed out, and it must not be affected by sunlight. Originally dyes were produced from plants and animals. In 1856 the English chemist, William Henry Perkin, created the first synthetic dye, mauve, from aniline made from benzene extracted from coal tar.

Coloured materials absorb some of the colours that make up white light and they reflect others. For example, a red dye reflects the red component of white light and absorbs the remaining colours of the spectrum, while a black dye absorbs all the colours. It has been found that certain groups of atoms in organic molecules are particularly good absorbers of certain colours. Such groups, called chromophores, are the nitroso group, $-N{=}O$, and the azo group, $-N{=}N{-}$. By varying the number and arrangement of the chromophore groups different colours can be obtained, making an immense colour range of synthetic dyes.

Fatty or oily dirt is the most difficult to remove. This kind of dirt forms a film on the surface of objects which water alone cannot

detergent action

1   garment with layer of dirt

2   garment immersed in water, molecules attracted to dirt

3   dirt lifted from surface of garment

4   molecules cover surface to prevent dirt settling

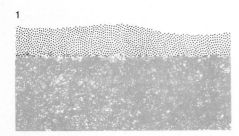

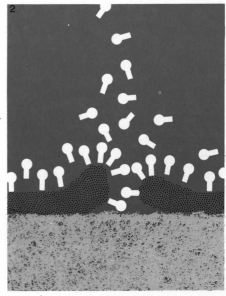

loosen. Soaps are long-chain hydrocarbon molecules with the group of atoms — COONa at one end. The hydrocarbon end is water-repellent (hydrophobic) while the — COONa group is attracted to water (hydrophilic). Fats and oils to be removed are attracted to the long-chain part of the soap molecule and are pulled into the water by the hydrophilic — COONa group. Hard soaps are usually sodium salts of stearic acid, $C_{17}H_{35}COOH$, palmitic acid, $C_{15}H_{31}COOH$, and oleic acid, $C_{17}H_{33}COOH$. The potassium salts of these acids are the basis of soft (liquid) soap.

Detergents are similar to soaps, but sulphur atoms replace the carbon atoms in the hydrophilic group. The dirt particles, surrounded by the detergent, enter the water as an emulsion, but do not, like soap, form a scum with hard water. Synthetic detergents consist of organic sulphates or sulphonates, produced from petroleum and sulphuric acid. However, commercial detergents also contain a surface-active material, a foaming agent to produce the lather and make the water look soapy, other chemicals to increase the cleaning efficiency and a substance to make the clothes appear whiter.

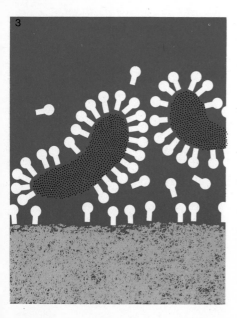

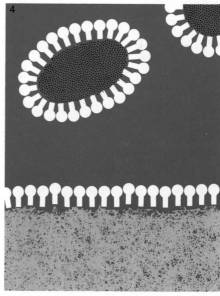

Drugs used in medicine are chemicals that affect the functioning of the human body and are given in controlled amounts to combat disease. There are many different types of medicinal drugs, and the treatment of diseases using these chemicals is called chemotherapy. An important group of drugs are those that kill or prevent the growth of the bacteria which cause infectious diseases. These drugs can be divided into three classes: antibiotics, disinfectants and antiseptics.

Antibiotics are substances that are produced naturally by various fungi, or can be synthesized chemically. Well-known examples are penicillin and streptomycin, which destroy the bacteria causing pneumonia and many other infections. Unfortunately, as new strains of bacteria which are resistant to these drugs are continually developing, there has to be a constant search for new forms of antibiotics to combat the new strains. Disinfectants are chemicals that kill bacteria and other harmful organisms outside the body. Antiseptics are drugs of synthetic origin which are also active against the bacteria that cause disease or putrefaction.

An important group of drugs used in chemotherapy are the

Right: *antibiotics, which come in many shapes and sizes, help doctors to fight disease*

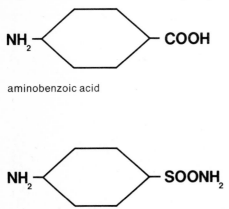

NH$_2$ ◁──────▷ COOH

aminobenzoic acid

NH$_2$ ◁──────▷ SOONH$_2$

sulphanilamide

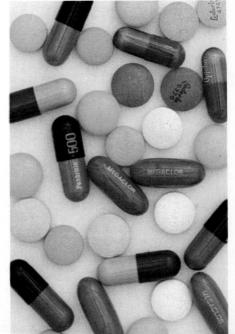

sulphonamides, some of which work by a competitive inhibition process. Certain disease bacteria require for their metabolism the chemical aminobenzoic acid, $NH_2C_6H_4COOH$, which they are able to find in small quantities in human blood. When sulphonamides are given to a patient the compound sulphanilamide, $NH_2C_6H_4SOONH_2$, is released into the bloodstream. The molecules of this chemical are shaped very like those of aminobenzoic acid and therefore the enzymes in the bacteria combine with the sulphanilamide instead of the aminobenzoic acid. Because they are deprived of aminobenzoic acid, which they need, the bacteria die.

Another important group of drugs are the anaesthetics, which cause loss of sensation. General anaesthetics, such as ether, Pentothal and cyclopropane, produce the total loss of consciousness needed for major operations. Local anaesthetics such as procaine make only a particular part of the body lose feeling. Another group of drugs that affect sensation are the analgesics, such as aspirin and phenacetin, which relieve pain. Narcotics, such as morphine (from opium), are stronger drugs, used to relieve severe pain, but they are addictive.

During surgery, anaesthetists rely on a variety of drugs to keep patients free from pain

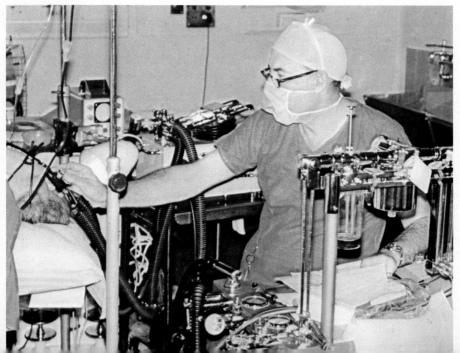

Chemistry
160   help for the farmer

A farmer needs fertile soil and a means of controlling weeds and preventing pests from damaging his crops. In the past, fields were kept fertile by resting them (leaving them fallow) once in three years and by rotation in which different crops were grown in a particular field each year. Weeds were kept down by hoeing but a farmer had almost no means of combating pests. Today chemists help the farmer to control weeds and eliminate pests.

The nitrogen necessary for plants to build up proteins comes from nitrogen compounds in the soil. In natural conditions the nitrogen compounds come from decay processes in the ground. Heavy cropping exhausts this supply so nitrogen must be added in the form of artificial fertilizers, which also provide the phosphorus and potassium needed for healthy plant growth. The main fertilizers are ammonium sulphate, potassium nitrate and calcium superphosphate. Blended with small quantities of other elements (trace elements) they form a compound fertilizer.

The contribution of the chemist

Hormones to combat plant pests are developed, prepared and tested in a laboratory

to weed control began with the discovery of selective weed-killers, chemicals that destroy weeds but do not harm the crops. They are closely related to plant hormones, the chemicals that regulate the growth of plants. An overdose of the particular hormone makes the weeds grow very rapidly and die. These hormones are complex organic chemicals, such as MCPA (methyl-chloro-phenoxyacetic acid).

Potato blight, caused by fungi, and tobacco mosaic and leaf-roll, caused by viruses, can be prevented by spraying. But the most serious menace to plants is insects which either eat them or spread diseases. Insecticides such as D.D.T. (dichloro-diphenyl-trichloroethane) and BHC (benzene hexachloride) can control many pests. But pesticides can destroy useful insects too, and kill wild birds that feed off the plants. Also, when the crops are eaten the poisonous chemicals can accumulate in the body. The natural environment is very sensitive to changes made by man, and when established food chains are broken, there may be unforeseen consequences. The importance of the study of ecology in relation to the use of insecticides and weed-killers has only recently been recognized.

Insecticides sprayed from helicopters are an important aid to modern farmers

Important dates in the development of the Physical Sciences

DATE	DISCOVERER	
5th century B.C.	Democritus	*first suggested the atomic structure of matter*
about 300 B.C.	Euclid	*elements of geometry*
3rd century B.C.	Archimedes	*laws of displacement of water*
about 230 B.C.	Eratosthenes	*first measurement of Earth's circumference*
3rd century B.C.	Aristarchus of Samos	*first suggested that Earth revolves around sun*
2nd century B.C.	Ptolemy	*celestial mechanics*
1543	Copernicus	*revival of Aristarchus' theory*
1559	Mercator	*first cylindrical map projection*
1589–92	Galileo Galilei	*the laws of falling bodies and the theoretical basis of heliocentric universe*
1600	William Gilbert	*discovery of properties of electric charge*
1609–19	Johann Kepler	*the laws of planetary motion*
1614	John Napier	*invention of logarithms*
1619	René Descartes	*invention of co-ordinate geometry*
1643	Evangelista Torricelli	*invention of the mercury barometer*
1662	Robert Boyle	*discovery of gas laws*
1670	Isaac Newton (Gottfried Leibniz independently 1675–6)	*discovery of calculus*
1676	Olaus Roemer	*first measurement of the velocity of light*
1687	Isaac Newton	*the laws of motion and gravitation*
1690	Christiaan Huygens	*wave theory of light*

DATE	DISCOVERER	
about 1789	Antoine Lavoisier	*discovery of the role of oxygen in combustion*
1799	Joseph Proust	*the law of definite proportions in chemistry*
1800	Alessandro Volta	*invention of the electric cell*
1803	John Dalton	*the Atomic Theory of Matter*
1820	Hans Christian Oersted	*discovery of electromagnetism*
1827	Georg Ohm	*discovery of Ohm's Law*
1828	Friedrich Wöhler	*first synthesis of urea*
1831	Michael Faraday	*discovery of electromagnetic induction*
1843	James Prescott Joule	*discovery of the mechanical equivalent of heat*
1850	Rudolph Clausius	*second law of thermodynamics*
1850s	Gustav Kirchhoff Robert Bunsen	*first spectroscopic investigations*
1854	George Boole	*invention of Boolean algebra*
1861	Friedrich Kekulé	*the structure of benzene*
1864	James Clerk Maxwell	*theory of electromagnetic radiation*
1869	Dmitri Mendeleyev	*periodic table of the elements*
1884	Svante Arrhenius	*discovery of ionic dissociation*
1887	Heinrich Hertz	*discovery of radio waves*
1895	Wilhemn Röntgen	*discovery of X-rays*
1896	Antoine Becquerel	*discovery of radioactivity*
1897	J. J. Thomson	*discovery of the electron*
1900	Max Planck	*first statement of quantum theory*

DATE	DISCOVERER	
1903	Ernest Rutherford	*discovery of the atomic nucleus*
1905	Albert Einstein	*special theory of relativity*
1911	C. T. R. Wilson	*invention of the cloud chamber*
1913	Niels Bohr	*Bohr model of the atom*
1913	Henry Moseley	*discovery of atomic numbers*
1915	Albert Einstein	*general theory of relativity*
1924–6	Louis de Broglie Erwin Schrödinger	*formulation of wave mechanics*
1925	Werner Heisenberg	*discovery of uncertainty principle*
1925	Wolfgang Pauli	*discovery of exclusion principle*
1932	James Chadwick	*discovery of the neutron*
1932	Carl D. Anderson	*discovery of the positron*
1938	Lise Meitner Otto Hahn	*discovery of nuclear fission*
1942	Enrico Fermi	*first nuclear pile*
1948	William Shockley	*invention of the transistor*
1957	USSR Academy of Science	*first man-made satellite orbits the Earth*
1958	James A. Van Allen	*Van Allen belts*
1960	Theodore H. Maiman	*invention of the laser*
1961	Yuri Gagarin	*first man in space*
1963	Thomas A. Matthews Allan R. Sandage	*discovery of quasars*
1968	Anthony Hewish	*discovery of pulsars*
1969	Neil Armstrong Edwin Aldrin Michael Collins	*first men on the moon*

Index

*Numbers in bold-face type indicate illustrations

Illustrations, acknowledgments and picture credits